Microturbines

Microturbines

Written and Edited by

Claire Soares, P.E.

AMSTERDAM • BOSTON • HEIDELBERG • LONDON
NEW YORK • OXFORD • PARIS • SAN DIEGO
SAN FRANCISCO • SINGAPORE • SYDNEY • TOKYO

Butterworth-Heinemann is an imprint of Elsevier

Academic Press is an imprint of Elsevier
30 Corporate Drive, Suite 400, Burlington, MA 01803, USA
525 B Street, Suite 1900, San Diego, California 92101-4495, USA
84 Theobald's Road, London WC1X 8RR, UK

Library of Congress Cataloging-in-Publication Data
Application submitted

British Library Cataloguing-in-Publication Data
A catalogue record for this book is available from the British Library.

ISBN 13: 978-0-7506-8469-9
ISBN 10: 0-7506-8469-0

For information on all Academic Press publications
visit our Web site at www.books.elsevier.com

Printed in the United States of America
07 08 09 10   9 8 7 6 5 4 3 2 1

Contents

Preface

The power industry is poised at the brink of major restructure. What used to be "large power" (i.e., massive corporations that produced power and sold it via international and national transmission networks) is now a swelling, increasingly mixed bag of players. The age of distributed power advances. None too soon, if one considers the current potential losses in power transmission (up to 30% of power generated in some less industrialized countries) and the huge amount of deforestation (with consequential increased greenhouse gas loads) that transmission lines often require.

Today's independent power producers, which may be consortiums of governments, private corporations, and original equipment manufacturers (OEMs) of power-generation machinery, grow in numbers and increasingly take over inefficient government-only power corporations. They may also construct and own their own transmission lines and distribution terminals.

Contemporary power producers today also include merchant power producers (MPPs). MPPs may own smaller generator units that are portable and can provide enough power for a mid-sized factory or small village. They may also operate stationary facilities for limited durations that are expected to yield adequate return on investment.

Just 30 years ago, "small power production" (SPP) was carried out mainly by massive process plants in locations so remote that conventional grids could not supply all their needs. Such plants, like the first Syncrude tar sands plant in northern Alberta (120,000 to 170,000 barrels of crude a day), produced most of its power consumption. The power machinery available at that time included gas and steam turbines that were about 20 megawatts (MW) in size, and that physically took up as much room as a 100-MW (and larger) gas turbine might today. Turbine technology, and especially controls technology, was still in its relative infancy. Similar small power production (other than that from small stand-alone systems like some wind turbines) usually required power lines to be brought to their doorstep for backup and peak power requirements.

However, in ice storms and hurricane seasons, downed power lines and poles cripple power reliability. Ultimately, everyone dreams of not being dependent on power lines of any sort or size. That day approaches—faster for those who practice distributed generation.

Meanwhile, *small power producers* (SPPs) today describes a quickly growing motley that includes large refineries, plastic conglomerates, and steel mills that may produce a waste fluid or flue gas that suffices as power-generation fuel. An SPP may be a village residents' co-op in Denmark that owns and shares the power from one 1.5-MW wind turbine. It may be a remote hospital or mid-sized factory that owns its own microturbine, which then allows them to be power independent. SPPs frequently maintain a grid interface to sell power back to that main grid when they produce in excess of their own requirements.

SPP equipment options are varied and several. However, the equipment item that is probably best suited in current day conditions, to extend the world of distributed (i.e., decentralized, non-megasized national company power model) power may be the microturbine. There are several reasons for this. The order of importance of those reasons is arguable, but they include the fact that as a small gas turbine, the microturbine is more like the conventional large turbines than say, wind turbines or solar cells. Further, they are not burdened with the "look nice, be acceptable to neighbors" requirements that frequently hold up solar cell household roof installation or massive wind farm developments.

These new trends are partially fostered by the failure of conventional "large power" to be absolutely reliable. In the United States and Canada particularly, the 1990s revealed a series of costly power industry Achilles' heels: crippling brown-outs in California (together with opportunistic power authorities in other states taking major monetary advantage of the situation), the collapse of the nuclear industry in Canada due in part to internal mismanagement, and failures of large sections of ice-laden power lines that cost severe financial loss in the northeastern United States. Consumers, government bodies such as the U.S. Department of Energy, and the research-and-development arms of major turbine manufacturers vowed that enough was enough.

And so we have today's quasi-distribution model that aims for a mix of large-scale conventional power, small and domestic industries distribution, as well as massive process plants and oil producers who make their own power. In addition, the electric power storage research community industry is pushing for a commercial debut with developments such as fuel cells. The power transmission line industry will take several decades to die, but with the exception of a few lines where no other technology makes sense, die it eventually shall.

Maximum turbine inlet temperature (TIT), the prime technology issue with large gas turbines, gave way with early microturbine applications to maximizing recuperator (waste heat recovery) performance. Simple microturbine applications can get up to 40% fuel efficiency with a recuperator. Once microturbines have their recuperator designs mastered, they then concentrate on the same issues as large gas turbines: maximizing TIT and pressure ratio (PR). However, in conjunction with a fuel cell, microturbine system efficiencies of 80% can be achieved. Large combined-cycle power-generation unit efficiencies still hover around 60%. So instead of a 300-MW combined-cycle power plant, a residential city can have several (joint or not) owners of microturbines with or without backup wind turbines, as well as households and building complexes with photovoltaic banks on their roofs: a distributed power scenario.

The current global political weather is affecting global fuels industries, as the current U.S. energy bill proves. The hype for fossil fuel needs strengthens, even as oil companies are allowed to increase stockpiles and charge the average consumer far more for gasoline. This continues, even as those same companies buy developed alternative energy technologies that save gasoline (such as higher-mileage cars), perhaps for the sole purpose of sitting on them until the windfall rise of gasoline prices has been exploited to its breaking point. The problem then is that the ramp-up to market of tested technology of this kind may take as long or longer than seven years.

Meanwhile, the "alternative energy" community, which includes all the major gas turbine and steam turbine manufacturers in the United States, are awarded Department of Energy grants to hone their alternative energy technologies. Sometimes they release them in a conservative or indirect manner that will not threaten their main product lines (the large turbines). They may, for instance, license microturbine packagers to sell microturbines that include a recuperator, the technology for which was developed by the OEM. Their research arm may have developed a fuel cell that they test in conjunction with a microturbine that they or a joint venture partner have developed.

The potential applications for microturbines, given their expanding partnership with fuel cells and a growing list of viable, unconventional fuels, are endless. All that said, if the general public do not educate themselves about new technologies, they will not know enough to pull growing market entry for alternative technologies out of OEMs and oil companies. Unless new items such as carbon dioxide legislation, tax credits for household SPPs, and tax rebates for business and household SPPs force OEMs and oil companies to promote smaller, distributed power, those big players will not push small power. So to not get swallowed up by rising fuel costs if you are an average person or small company, it's pull not push time. Unfortunately, most of the U.S. public does not have a history of pulling. However, knowledge has its own powerful pull. So if you pick up this book and check out even some of what's in it, you're pulling.

Author's Notes

- Microturbines, fuel cells, and hybrids are relatively new technologies. They all support distributed power generation and greenhouse gas emission reduction in some way, so technological development work in these areas is heavily subsidized by state and federal governments.
- Therefore, most of the references in this book are extracts of government or government supported studies. Unlike with more mature fields, such as pumps and compressors, microturbine manufacturers have not yet built up their own financially independent R&D and commercial literature departments. So the government or government-supported studies are the best references for a basic understanding of microturbines, fuel cells, and hybrids. The owners of the referenced studies (such as NREL) maintain their own copyright of their original documentation, and reprints (with permission) of extracts or adapted extracts of their work in this book ought not to be construed as infringing on that copyright.
- URL internet addresses are always a moving target, as most web administrators update their sites often. Also, companies change hands and mergers happen. So, if a quoted URL address is not as mentioned in the book, first go to the root website, and then use the search function to find what you need.
- Cost data is also a constantly moving target due to fuel prices and major upheavals, such as wars and currency devaluations. So consider all cost data in this book as good for comparing the relative costs of different technologies, but not in absolute terms (although it may have been current at the time of writing).
- As this is a first edition, the manuscript is smaller than the author would have liked. However, much of the material which fossil-fueled turbine owners will need (regardless of whether they operate a microturbine or a 600-megawatt combined cycle plant), is in *Gas Turbines: A Handbook of Air, Land, and Sea Applications* by Claire Soares, also published by Butterworth-Heinemann.

Introduction and Background Glossary

This introductory chapter provides background detail and a glossary of terms for the layperson or professional in an unrelated field. This will help them grasp the basics of microturbine and microturbine system(s) (i.e., microturbine[s] in conjunction with other equipment, such as a fuel cell or diesel generator) technology. The information may be presented in essay or question-and-answer type format. If the term sought is not found here, check the index.

For readers who are relatively new to this field, proceeding to Part 2 first would be useful, as this gives a broad idea of potential systems applications, technology to date, relevant state of legislation, and so forth. It also gives an idea of what to look for if the reader is considering business investment in a microturbine company (however, this book does not attempt to give legal or financial advice). It therefore is also useful to investment consultants and management accountants who have to study portfolios that relate to this subject.

The chapters in Part 1 detail information of significance to the person, often an engineering professional, who has or plans to operate microturbines in his or her facility or plant.

The entire book has been written and edited so that it is within the grasp of good high school students. It also provides enough detail for university students planning to work in this or related fields, such as rotating machinery manufacture and service, power generation, and process industries.

GLOSSARY

Absorption chiller: Water chiller based on absorption of refrigerant vapor into a liquid solution, pumping of solution to elevated pressure, and desorption of refrigerant vapor through addition of heat. Direct-fired chillers employ natural gas burners, indirect-fired chillers use steam or hot water from a boiler, heat-recovery heat exchanger, or generator exhaust gas. Single-, double-, and triple-effect chillers employ multiple stages of desorption and internal use of waste heat to boost efficiency.

Demand charge: Charges for the use of electricity based on the maximum power requirement, electrical demand, during a specified period of time, typically a month (dollars per kilowatt [kW]).

Desiccant: A solid or liquid material with an affinity for absorbing water molecules.

Engine-generator: Electrical generator using a reciprocating, Stirling, or rotary engine.

Enthalpy wheel: Heat exchanger rotating through building supply and exhaust airflows to transfer energy from one air stream to the other.

Evaporative cooling: Lowering the temperature of air through the evaporation from a water spray or wetted membrane. Direct evaporative cooling adds water to the supply air while indirect evaporative cooling adds water to the exhaust air and incorporates a heat pipe or thermal wheel for indirect cooling of the supply air.

Fuel cell: Device for producing electricity using a chemical process rather than conventional combustion processes with steam generators.

Heat wheel: Heat exchanger rotating through building supply and exhaust air flows to transfer heat from one air stream to another.

Latent cooling load: Amount of cooling required to reduce humidity of air in conditioned space to specified level for comfort.

Microturbine power generator: Turbine-engine–driven electrical generator with output power of 25 to 300 kW.

Teal-time pricing: Charges for electrical demand and consumption based on instantaneous cost of production and distribution as opposed to fixed rates or fixed time-of-day rates.

Sensible cooling load: Amount of cooling required to reduce the temperature of air in the conditioned space to a specified level for comfort.

Therm: A unit of energy used to measure natural gas; 10^5 BTU, 1.055×10^8 joules, approximately 97.5 ft^3.

Thermal wheel: Heat exchanger rotating between two airflows to transfer heat from one to the other.

Ton or refrigeration ton: Quantity of cooling available from melting 2000 pounds of ice; 12,000 BTU/hour or 3.1413 kW.

Vapor compression air conditioning: Cooling system based on mechanical compression of a gaseous refrigerant to a high pressure, and heat transfer with changes of state (e.g., liquid and vapor) to produce useful heating or cooling.

Waste heat: Portion of the energy input to a mechanical process that is rejected to the environment.

BATTERY, ELECTRIC[*]

A *battery* is a device that converts chemical energy into electrical energy, consisting of a group of electric cells that are connected and act as a source of direct current (DC). The term is also now commonly used for a single cell, such as the alkaline dry cell used in flashlights and portable tape players, but strictly speaking batteries are made up of connected cells encased in a container and fitted with terminals to provide a source of direct electric current at a given voltage.

A cell consists of two dissimilar substances, a positive electrode and a negative electrode, that conduct electricity, and a third substance, an electrolyte, that acts chemically on the electrodes. The two electrodes are connected by an external circuit (e.g., a piece of copper wire); the electrolyte functions as an ionic conductor for the transfer of the electrons between the electrodes. The voltage, or electromotive force, depends on the chemical properties of the substances used but is not affected by the size of the electrodes or the amount of electrolyte.

---

[*] Source: Adapted from Columbia University Press Encyclopedia.

Batteries are classified as either dry cell or wet cell. In a dry cell the electrolyte is absorbed in a porous medium or is otherwise restrained from flowing. In a wet cell the electrolyte is in liquid form and free to flow and move. Batteries also can be generally divided into two main types:

- Rechargeable
- Disposable

Disposable batteries, also called *primary cells*, can be used until the chemical changes that induce the electrical current supply are complete, at which point the battery is useless. Disposable batteries are most commonly used in smaller, portable devices that are only used intermittently or at a large distance from an alternative power source or have a low current drain.

Rechargeable batteries, also called *secondary cells*, can be reused after being drained. This is done by applying an external electrical current, which causes the chemical changes that occur in use to be reversed. The external devices that supply the appropriate current are called *chargers* or *rechargers*.

A battery called the *storage battery* is generally of the wet-cell type; i.e., it uses a liquid electrolyte and can be recharged many times. The storage battery consists of several cells connected in series. Each cell contains a number of alternately positive and negative plates separated by the liquid electrolyte. The positive plates of the cell are connected to form the positive electrode; similarly, the negative plates form the negative electrode.

In the process of charging, the cell is made to operate in reverse of its discharging operation. In other words, current moves through the cell in the opposite direction, causing the reverse of the chemical reaction that ordinarily takes place during discharge, so that electrical energy is converted into stored chemical energy.

The storage battery's greatest use has been in the automobile, where it was used to start the internal-combustion engine. Improvements in battery technology have resulted in vehicles—some in commercial use—in which the battery system supplies power to electric drive motors instead. Electric cars have not made a large dent in the consumer market yet. One issue they face is the matter of battery disposal if the battery contains lead. Electric cars give an illusion of "green energy." Some forget that the power used to charge their batteries generally comes from conventional fossil fuels.

Batteries are made of a wide variety of electrodes and electrolytes to serve a wide variety of uses. Batteries consisting of carbon-zinc dry cells connected in various ways (as well as batteries consisting of other types of dry cells) are used to power such devices as flashlights, lanterns, and pocket-sized radios and CD players.

Alkaline dry cells are an efficient battery type that is both economical and reliable. In alkaline batteries, the hydrous alkaline solution is used as an electrolyte; the dry cell lasts much longer as the zinc anode corrodes less rapidly under basic conditions than under acidic conditions.

In the United States, the lead storage battery is commonly used. A more expensive type of lead-acid battery called a *gel battery* (or gel cell) contains a semisolid electrolyte to prevent spillage. More portable rechargeable batteries include several dry-cell types, which are sealed units and are therefore useful in appliances like mobile phones and laptops. Cells of this type (in order of increasing power density and cost) include nickel-cadmium (nicad or NiCd), nickel metal hydride (NiMH), and lithium-ion (Li-Ion) cells.

Evidence suggests that primitive batteries were used in Iraq and Egypt as early as 200 BC for electroplating and precious metal gilding. In 1748, Benjamin Franklin coined the term *battery* to describe an array of charged glass plates. However, most historians date the invention of batteries to about 1800, when experiments by Alessandro

Volta resulted in the generation of electrical current from chemical reactions between dissimilar metals.

Experiments with different combinations of metals and electrolytes continued over the next 60 years. In the 1860s, Georges Leclanche of France developed a carbon-zinc wet cell; non-rechargeable, it was rugged, manufactured easily, and had a reasonable shelf life. Also in the 1860s, Raymond Gaston Plant invented the lead-acid battery. It had a short shelf life, and around 1881 Émile Alphonse Faure developed batteries using a mixture of lead oxides for the positive plate electrolyte with faster reactions and higher efficiency.

In 1900, Thomas Alva Edison developed the nickel storage battery, and in 1905 the nickel-iron battery. During World War II the mercury cell was produced. The small alkaline battery was introduced in 1949. In the 1950s, the improved alkaline-manganese battery was developed.

In 1954 the first solar battery or solar cell was invented. In 1956 the hydrogen-oxygen fuel cell was introduced. The 1960s saw the invention of the gel-type electrolyte lead-acid battery. Lithium-ion batteries, wafer thin and powering portable computers, cell phones, and space probes were introduced in the 1990s. Computer chips and sensors now help prolong battery life and speed the charging cycle. Sensors monitor the temperature inside a battery as chemical reactions during the recharging cause it to heat up; microchips control the power flow during recharging so that current flows in rapidly when the batteries are drained and then increasingly slowly as the batteries become fully charged.

Another source of technical progress is nanotechnology. Research indicates that batteries employing carbon nanotubes will have twice the life of traditional batteries.

FUEL CELL[*]

For details on the working of a fuel cell, different types of fuel cells, their use in conjunction with microturbines, case histories, and other details, see Part 2 of this book.

Q1: Why have fuel cells received attention in the popular press in energy-conscious states like California recently?

A: Fuel cells offer good energy efficiency and reduced environmental impact.

Q2: How will the NFCRC contribute to the commercialization of fuel cell technology?

A: The NFCRC will conduct research for commercialization, educate students, demonstrate fuel cell technology to industry and the public, and promote technology and information transfer from universities and industry to the general public.

Q3: In what time frame should we expect fuel cells to affect the lifestyle of a typical U.S. citizen?

A: Consumers of specialized products and vehicles could be affected within the next 5 years; the majority of the general public will begin enjoying the benefits of more widespread fuel cell application within the next decade.

---

[*] Sources: Adapted from information from the National Fuel Cell Research Center (NFCRC), University of California–Irvine; and Soares, C. 2005. *International Power Generation*.

Q4: Are fuel cells available in the market today?

A: There are several manufacturers of fuel cells that will sell integrated fuel cell power plants either today or in the very near future. These manufacturers include many NFCRC members with most selling only "demonstration" plants at this time.

ONSI Corporation, a subsidiary of International Fuel Cells, sells the only commercial fuel cell power plant in the world—the 200-kW PC25 power plant. A few units are at approximately 5 years of commercial operation.

The longest sustained operation has been achieved by a unit located at the Hyatt Regency in Irvine, CA. The longest continuous operation without an outage exceeds 1 year and has been achieved by a unit operated by Tokyo Gas.

Q5: For what reasons are people currently using fuel cells?

A: Most of the current electricity production from fuel cells is being used in stationary power applications like providing power to small industrial sites, hospitals, hotels, etc. Other applications include space applications (e.g., space shuttle), transportation demonstrations (e.g., buses, automobiles), and portable power applications (e.g., portable computers, communications equipment).

Q6: What applications for fuel cells exist in alternative energy-conscious states like California?

A: Current fuel cell applications in California include but are not limited to:

- A Westinghouse 25-kW tubular solid oxide fuel cell (SOFC) has been installed at the National Fuel Cell Research Center.
- Eight ONSI 200-kW phosphoric acid fuel cells (PAFCs) are installed at the South Coast Air Quality Management District, the Irvine Hyatt Hotel, Kraft Foods, Kaiser Hospital in Anaheim and Riverside, University of California, Santa Barbara, Santa Barbara Jail, and Vandenberg Air Force Base.
- M-C Power Corporation has installed two 250-kW molten carbonate fuel cells (MCFCs) at the Unocal Research Center in Brea and the Marine Air Station Miramar in San Diego.
- Energy Research Corporation's 2.0-megawatt [MW] MCFC is installed at the City of Santa Clara, in Santa Clara.

Q7: What potential future applications are likely in states like California?

A: Future applications of fuel cells in California will likely include a large fraction of stationary distributed power-generation units that will become increasingly utilized in locations where grid-supplied power becomes costly, due to either grid overloading or remote location.

Fuel cells could also play a major role in reducing emissions if widely applied in automobiles. Finally, individual California residents may be able to purchase fuel cells for their homes in the near future, to provide clean, efficient, and reliable electricity from natural gas already supplied to the home.

Q8: What future applications are likely in the U.S. and international markets?

A: The international market for power generation is much larger than the domestic market. This is due primarily to the fact that most countries (in particular, developing countries) do not have the comprehensive utility grid of the United States.

Many countries are attempting to get electricity to their more remote areas and increase their industrial output requiring increased power generation. In addition, since these countries do not have an extensive electrical grid, distributed power generation (to which fuel cells are especially well suited) may play a more important role. Transmission and distribution line costs can be reduced in much the same manner as cellular telephone systems, which bypass the installation of telephone lines and are surpassing the use of traditional phone systems in many developing countries.

Q9: What fuel cell technologies do you see playing a role?

A: Each of the fuel cell types currently under development or manufacture has features that make it particularly attractive to use in certain applications. For example, the low-temperature operation and high-power density of proton exchange membrane fuel cells (PEMFCs) make them well-suited for automotive application.

However, the increased efficiency and higher-temperature operation of either MCFCs or SOFCs make them more amenable to stationary power and/or hybrid applications with a gas turbine engine. Therefore, it would be premature to neglect developing each of the fuel cell types currently under consideration.

Q10: Why are fuel cells more efficient at converting fossil fuel to electricity than conventional heat engines?

A: Fuel cells convert fossil fuel energy directly to electricity, whereas heat engines first convert the chemical energy to thermal energy, then to mechanical energy, and finally to electrical energy. Even though the fuel cell can only convert 50% to 60% of the fuel chemical energy to electricity, this is considerably higher than efficiencies with heat engines.

The fuel cell, because it does not use a thermal energy step, is not subject to the Carnot cycle limits that are imposed on heat engines.

Q11: Why do fuel cells exhibit efficiency improvements with increasing operating pressure?

A: The dependence of the reversible cell potential is given by the Nernst equation, which describes mathematically the theoretical operation of a fuel cell. This equation shows that as the operating pressure increases the reversible cell potential or voltage also increases. The rate of cell potential increase declines at high pressures.

Enhanced cell voltages are due largely to the increase in reactant partial pressures and the resulting improvements in mass transport rates. This has led to the use of hybrid systems with gas turbines and fuel cells together. The gas turbine provides a high-pressure environment for the fuel cell while using the fuel cell waste thermal energy as its energy source.

Q12: Over the last 30 years, fuel cells have been funded on the basis of the promise that they are a product everyone will have in their backyard within the next few years producing cheap electricity and water. What happened?

A: First, note that there are many different kinds of fuel cells just as there are many types of heat engine, such as diesel engines and gas turbines. By analogy, the funding has been applied to all "heat engines," not to any specific example. Thus, the funding received by a particular fuel cell type is a small fraction of that reported.

Some fuel cell types are presently commercially viable in special niche markets and, as production costs decrease, the more advanced fuel cells will move into widespread use. However, it is unlikely that "everyone will have a fuel cell in their backyard for personal use." Additionally, some start-up fuel cell companies have been guilty of exaggerated claims.

As the industry grows, more realistic claims and projections are being made as to the commercial viability of the products being offered. Finally, funding for other power generation and power technologies is far more than that awarded to fuel cell technologies, primarily because fuel cells have not been seriously considered for military and strategic applications.

Q13: What distinguishes different types of fuel cells?

A: Fuel cells are generally characterized by the type of electrolyte used. The primary types of fuel cells include the following (in alphabetical order):

1. Alkaline fuel cells (AFCs), which contain a liquid alkaline electrolyte
2. MCFCs, which contain a molten carbonate salt electrolyte at operating temperatures ($\sim$650°C)
3. PAFCs, which contain a phosphoric acid electrolyte
4. PEMFCs, which contain a solid polymer electrolyte
5. SOFCs, which contain a solid ceramic electrolyte

Each of these fuel cell types operates on the basic principles of reverse electrolysis.

Q14: What fuel does a fuel cell use?

A: All fuel cells utilize hydrogen as the basic fuel for the reverse hydrolysis process, converting the energy contained in the hydrogen fuel directly to electricity while forming water as the end product. The source of the hydrogen may be a fossil fuel, such as natural gas or gasoline; in this case, the fuel must be reformed to produce hydrogen.

Some fuel cell types (e.g., SOFC and MCFC) can accept a carbon monoxide–hydrogen mixture that can be produced from readily available fuels. These fuel cells also convert carbon monoxide into carbon dioxide very effectively. Other fuel cell types (e.g., PAFCs, PEMFCs) require a relatively pure hydrogen stream as the main fuel for the fuel cell. This involves additional processing of fossil fuels.

Q15: In addition to the fuel cell power plant itself, what technology developments are needed in order to apply fuel cells in stationary power applications?

A: Stationary power applications of fuel cells will require the development of a power inverter and grid interface technology that is efficient and cost effective. Power inversion is required to convert the direct current (DC) power produced by the fuel cell stack to the alternating current (AC) of the utility grid. In addition, this grid interface technology must account for phase and frequency matching of the AC produced by the fuel cell with that of the grid.

Controls technology for reliable and cost-effective operation of fuel cells and accounting for the power production and power quality of the fuel cells system is needed. In addition, reforming technologies for the conversion of readily available fuels to hydrogen will be required. This reformer technology will most likely need to reform natural gas as the first fuel of choice.

Q16: What technology developments are needed in order to apply fuel cells in transportation applications?

A: Arguably the most important developments required for transportation applications of fuel cells involve fuel handling and fuel processing. Since the fuel cell considered the primary candidate for transportation applications, the PEMFC, requires a clean hydrogen fuel, stringent requirements are placed upon the processing of typical transportation fuels like gasoline and methanol. The development of compact, efficient, cost-effective, and high-purity hydrogen-producing reformer technology is a key requirement.

Another strategy that could avoid the "on-board" reformation of liquid transportation fuels is the storage and direct use of hydrogen. This would require significant advances in the storage of hydrogen. This may be done with metal hydride or carbon nanotube storage of hydrogen, or the development of crash-worthy high-pressure hydrogen-storage technology. It would also require significant development of hydrogen supply infrastructure. The former approach would use more widely available transportation fuels.

Q17: What are the principal barriers to the development of commercial fuel cell products?

A: The principal barrier to commercial fuel cell products is manufacturing costs. Today, all fuel cell products cost more to manufacture than similar, already available products. This is due to several factors, including the following:

1. No economies of scale
2. No economies of volume
3. Fuel cell manufacture to date has typically been accomplished in laboratories, not in manufacturing plants
4. The complexity of fuel cell systems (despite the fact that fuel cells themselves are simple)
5. The high cost of materials (e.g., precious metals) needed to make fuel cells
6. Manufacturing experience
7. Optimized manufacturing techniques

Another barrier is fuel flexibility. Fuel cells operate optimally on hydrogen. However, they must be able to use readily available hydrocarbon fuels before they can be viable commercial products. The lack of a history of widespread use or general public acceptance of fuel cells are other barriers.

Q18: What technical issues must be addressed to overcome barriers to commercialization of fuel cells?

A: Important technical issues that need to be addressed include the development of fuel cell products that use lower-cost materials, contaminant-tolerant materials, easily manufactured materials, and materials amenable to fuel flexibility. Further, manufacturing processes must be developed that will allow inexpensive, high-volume manufacturing of fuel cell products.

It is expected that fuel cell systems will be constructed at sizes that would allow manufacturing cost reduction through economies of volume (not economies of scale). Innovative reformers and/or innovations that reduce the cost of traditional fuel-reformation technologies are needed for fuel flexibility.

In addition, the advancement of fuel cell systems (including high-efficiency fuel cells, cogeneration technologies, and hybrid fuel cell heat engine technologies) is needed. Lastly, the majority of plant items such as pumps, valves, piping, controls, and power electronics (e.g., inverters) need improvements in reliability, cost, and optimization for fuel cell applications.

Q19: What other issues could affect the full commercialization of fuel cell products?

A: Other issues affecting fuel cell commercialization include yet-to-be-determined rules and regulations regarding siting, insuring, and certifying fuel cell products. Also, business issues—such as the depreciation rate allowed to those who purchase fuel cell products and the manner in which banks lend money for purchasing fuel cells—will affect the introduction of fuel cell products.

Regulatory issues concerning pollutants, such as nitrogen oxides, carbon monoxide, and hydrocarbons, could be made more restrictive and thereby facilitate installation and use of more fuel cells. Another significant boost for fuel cells' entry into the marketplace could be credits and/or financial reward for the aversion, limitation, and/or reduction of global climate change gases, such as carbon dioxide.

Q20: What is a hybrid fuel cell system, and what are its advantages over other advanced energy-production plants?

A: Hybrid technology represents the union of two separate power plants carefully developed and integrated into a single unit. The union creates a synergy between the power systems, exploiting the benefits of each. This advantage allows for improved overall performance. For example, the NCFRC will soon acquire a hybrid SOFC/microturbine generator system.

This plant will use hot fuel cell exhaust gas to drive a microturbine, which will then provide a high-pressure environment to enhance the electrical output of the fuel cell. This system will operate more economically than either an SOFC or microturbine generator running alone. With most fuel cell hybrids, the advantage over other advanced power-generation technologies is fuel efficiency. Instead of being vented into the environment as waste heat or used for cogeneration, the fuel cell's exhaust is captured and its energy utilized for the production of electricity enhancing the overall system efficiency.

Q21: In addition to microturbine generators, what other types of power devices could be used to produce a hybrid fuel cell system?

A: Hybrid systems that include combinations of fuel cells with conventional power production, power storage, energy conversion, or energy management devices are extremely promising. Heat engines, like microturbine generators, reciprocating or Stirling engines, could run using a fuel cell's waste heat to produce additional electricity and/or pressurize the fuel cell. Batteries and flywheels could be used to store power and then feed it back to the system when energy demand becomes high. Fuel cells can also run with other fuel cells for greater hydrogen production and chemical energy storage.

Q22: What do you think is the key challenge for the future with bringing fuel cells into commercial use?

A: The key challenge will be to figure out a way to allow the ideal hydrogen-fueled engine (a fuel cell) to cost-effectively produce power in the hydrocarbon-based economy that we live in today. This is the most significant challenge with regard to integrating fuel cells with available infrastructure, reducing the capital cost of fuel cell systems through volume manufacturing, and achieving widespread use in various sectors.

Technical hurdles vary depending upon the fuel cell application of interest. For example, automotive applications will need to determine how best to use the available infrastructure for the production, distribution, delivery, and storage of liquid fuels for fuel cell automobiles. This could involve everything from a cost-effective on-board gasoline reformer, to innovative hydrogen storage technology, to replacement of the entire infrastructure.

A second example is that of distributed power generation, where high-temperature fuel cells need to reduce the capital cost of the fuel cell system to allow them to compete with increasingly efficient heat engines operated on inexpensive natural gas. A third example is that of battery replacement where hydrogen storage using innovative, safe, and high-storage-density technology will need to be developed for widespread use of fuel cells in place of batteries.

Q23: Can ethanol be used as a fuel cell fuel?

A: Ethanol (or ethyl alcohol) is a viable fuel for fuel cells. Ethanol can be reformed to produce a hydrogen-rich gas stream that can be supplied to an SOFC or MCFC directly. Removal of the carbon monoxide in the hydrogen-rich gas would be required before it could be used to fuel other fuel cell types.

Ethanol reformation is easier than that of gasoline, diesel, or other distillate fuels since it is a pure compound and can be reformed at lower temperatures. Ethanol is more difficult to reform than methanol (a primary candidate fuel for transportation applications of fuel cells).

Also, ethanol has a lower hydrogen-to-carbon (H/C) ratio compared with methanol. Ethanol, CH_3CH_2OH, has an H/C ratio of three to one, while that of methanol, CH_3OH, is four to one. The final viability of ethanol versus methanol, gasoline, or other liquid fuel for use in fuel cells will be determined by the market price that the fuel can deliver and the useable hydrogen reformate it can deliver.

At this time (last quarter 2005, U.S. prices), both ethanol and methanol are more expensive to produce than gasoline. Some argue that the lower cost of gasoline is due to hidden external subsidies to the oil industry (for instance, International Center for Technology Assessment. December 1998. *The real price of gasoline*. See http://www.ethanol.org.).

The politics of oil and conventional gasoline is a subject that may be determined, to a large extent, by wars and the threat of war. The war in Iraq and the current situation with Iran's stand-off with regard to its development of nuclear technology are good examples.

In 10 years, about 25% of the U.S. oil supply is likely to come from Africa, from fields such as those off the coast of Angola. Another 25% can potentially come from the deposits in Azerbaijan, which the West hopes to receive via a pipeline being constructed through Azerbaijan, Georgia, and Turkey. These developments could alter the face of the entire demand/price structure of oil. Under those circumstances, alternative fuels could become less attractive, if all supplies of "new" oil remain uninterrupted.

Q24: If everything in my house, including my car, were run off fuel cells today, how much money would I save on my energy bills annually?

A: With the high initial cost of fuel cell systems today, and the amortization of these capital costs spread over a 10-year period, the average consumer would break even with regard to annual energy costs if they were currently paying around $0.10 to $0.14/kWh for electricity. The average rate in the U.S. today is around $0.09 to $0.13 per kWh, but there are many markets (e.g., remote locations, islands) that pay more for electricity.

The capital cost of a fuel cell system may drop dramatically in the upcoming years, depending on research and manufacture. Then cost savings could occur. For example, if fuel cell system capital cost was consistently reduced to $1000/kW installed, then electricity could be produced at less than $0.04/kWh, in which case the electricity energy bills could be cut roughly in half compared with current average U.S. rates. These figures assume a constant cost for natural gas and a 10-year amortization of capital costs.

The energy costs that could result from efficiency increases in an automobile equipped with a fuel cell engine at the same price as an internal combustion engine are also roughly half of the current costs. This is due to the projection that fuel cell electric vehicles would obtain around 80 miles/gallon of gasoline equivalent compared with around 40 miles/gallon for advanced internal combustion engine concepts. The hybrid internal combustion engine–electric vehicle concept could also achieve energy costs of approximately one half due to similar projections for gas mileage (~70 miles/gallon).

That said, the consumer drives the market wheels that determine figures such as gas mileage. A conventional Honda Accord (old model, 135-horsepower engine) can get between 34 and 40 miles/gallon, depending on how the engine is tuned and run. A new Honda hybrid Accord claims about 40 miles/gallon. The reason? The engine was enlarged to about 250 horsepower. The reason for that, in turn, is consumer demand.

With an American consumer who wants to feel "safe" in a post-"September 11th" world a gas-guzzling Hummer or SUV has enough of a military tank feel to provide that comfort. If the consumer does not want an automobile that looks like a tank, then speed (horsepower) is another comfort-providing feature.

In other words, even when the energy-savings technology is available on a consumer scale, the American consumer may choose comfort and safety instead. That said, new energy technologies are flourishing in Europe. More crowded streets, fewer natural oil and gas resources, and a totally different attitude to life may yet make Europe the continent where many alternative technologies first reach their full potential.

Q25: What happens if you're driving an electric or fuel cell electric vehicle and it breaks down in the middle of the road?

A: A breakdown in an electric vehicle or fuel cell electric vehicle would not be much different than that in a current internal combustion engine–driven vehicle. The breakdown, in most cases, would be expected to occur less frequently and be less expensive.

For example, radiator system leaks (which can spew steam into the air) would not occur with an electric vehicle (or fuel cell electric vehicle) that does not contain a radiator. In addition, the systems involved in an electric vehicle are less complex and involve fewer parts or systems, so the breakdown would be expected to be less frequent.

For fuel cell electric vehicles, this is not necessarily the case. The repair of an electric or fuel cell electric vehicle is not readily accomplished by the typical mechanic, so one might in the short term use a certified mechanic from the dealer and use a roadside assistance program (provided by most electric vehicle and fuel cell electric vehicle manufacturers today). In the long term, one can expect to have a wide array of skilled mechanics to repair electric vehicles and fuel cell electric vehicles.

GAS TURBINE[*]

A *gas turbine* is a rotating engine that extracts energy from a flow of combustion gases that result from the ignition of compressed air and a fuel (either a gas or liquid, most

---

[*] Sources: Adapted from Bloch, H. P. and Soares, C. 1998. *Process plant machinery,* 2nd ed. Butterworth-Heinemann. Boston, MA; Soares, C. 2007. *Gas turbines: a handbook of air, land, and sea applications.* Butterworth-Heinemann. Boston, MA; and Claire Soares's personal notes from ASME IGTI annual conference proceedings 1985 to 2003.

commonly natural gas). It has an upstream compressor module coupled to a downstream turbine module, and a combustion chamber(s) module (with igniter[s]) in between. (*Gas turbine* may also refer to just the turbine module.)

Energy is added to the gas stream in the combustor, where air is mixed with fuel and ignited. Combustion increases the temperature, velocity, and volume of the gas flow. This is directed through a diffuser (nozzle) over the turbine's blades, spinning the turbine and powering the compressor.

Energy is extracted in the form of shaft power, compressed air, and thrust, in any combination, and used to power aircraft, trains, ships, generators, and even tanks.

The Gas Turbine Cycle

The simplest gas turbine follows the Brayton cycle (Figure 1). In a closed cycle (i.e., the working fluid is not released to the atmosphere), air is compressed isentropically, combustion occurs at constant pressure, and expansion over the turbine occurs isentropically back to the starting pressure.

As with all heat engine cycles, higher combustion temperature (the common industry reference is *turbine inlet temperature*) means greater efficiency. The limiting factor is the ability of the steel, ceramic, or other materials that make up the engine to withstand heat and pressure.

Considerable design/manufacturing engineering goes into keeping the turbine parts cool. Most turbines also try to recover exhaust heat, which otherwise is wasted energy. Recuperators are heat exchangers that pass exhaust heat to the compressed air, prior to combustion. Combined-cycle designs pass waste heat to steam turbine systems, and combined heat and power (i.e., cogeneration) uses waste heat for hot water production.

Mechanically, gas turbines can be considerably less complex than internal combustion piston engines. Simple turbines might have one moving part: the shaft/compressor/turbine/alternator-rotor assembly, not counting the fuel system.

More sophisticated turbines may have multiple shafts (spools), hundreds of turbine blades, movable stator blades, and a vast system of complex piping, combustors, and heat exchangers.

The largest gas turbines operate at 3000 (50 hertz [Hz], European and Asian power supply) or 3600 (60 Hz, U.S. power supply) RPM to match the AC power grid. They

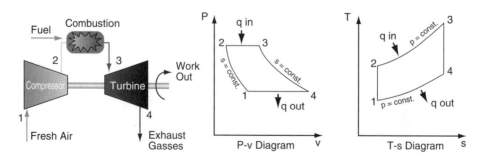

FIGURE 1   *Idealized Brayton cycle. (Source: Wikipedia.)*

require their own building and several more to house support and auxiliary equipment, such as cooling towers.

Smaller turbines, with fewer compressor/turbine stages, spin faster. Jet engines operate around 10,000 RPM and microturbines around 100,000 RPM.

Thrust bearings and journal bearings are a critical part of the design. Traditionally, they have been hydrodynamic oil bearings or oil-cooled ball bearings. This is giving way to hydrodynamic foil bearings, which have become commonplace in microturbines and auxiliary power units (APUs).

Gas Turbines for Electrical Power Production

The General Electric H series power-generation gas turbine is pictured in Figure 2. This 400-MW unit has an official rated thermal efficiency of 60% in combined-cycle configurations.

Power plant gas turbines range in size from truck-mounted mobile plants to enormous, complex systems.

They can be particularly efficient—up to 60%—when waste heat from the gas turbine is recovered by a conventional steam turbine in a combined-cycle configuration.

Simple-cycle gas turbines in the power industry require smaller capital investment than combined-cycle gas, coal, or nuclear plants and can be designed to generate small or large amounts of power. Also, the actual construction process can take a little as several weeks to a few months, compared to years for baseload plants. Their other main advantage is the ability to be turned on and off within minutes, supplying power during peak demand. Large simple-cycle gas turbines may produce several hundred megawatts of power and approach 40% thermal efficiency.

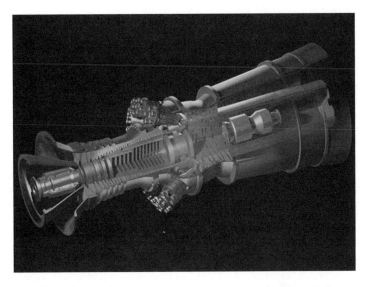

FIGURE 2 *General Electric H series electric power-generation gas turbine. (Source: U.S. Department of Energy.)*

Auxiliary Power Units

APUs are small gas turbines designed for auxiliary power of larger gas turbines, usually aircraft. They are frequently used to start the larger gas turbine. A small electric motor provides the APU with the initial air intake so it can start itself.

These units are well suited for supplying compressed air for aircraft ventilation (with an appropriate compressor design), start-up power for larger jet engines, and electrical and hydraulic power. (These are not to be confused with the auxiliary propulsion units, also abbreviated *APUs,* aboard the gas turbine–powered Oliver Hazard Perry-class guided-missile frigates. The Perrys' APUs are large electric motors that provide maneuvering help in close waters or emergency backup if the gas turbines are not working.)

Gas Turbines in Vehicles

Gas turbines are used on ships, locomotives, helicopters, and in the M1 Abrams and T-80 tanks. A number of experiments have been conducted with gas turbine–powered automobiles.

In 1950, designer F. R. Bell and Chief Engineer Maurice Wilks from British car manufacturer Rover unveiled the first car powered with a gas turbine engine. The two-seater JET1 had the engine positioned behind the seats, air intake grills on either side of the car, and exhaust outlets on the top of the tail. During tests, the car reached top speeds of 140 kilometers (km)/hour, at a turbine speed of 50,000 RPM. The car ran on petrol, paraffin, or diesel oil, but fuel-consumption problems proved insurmountable for a production car. It is currently on display at the London Science Museum.

Rover and the BRM Formula One team joined forces to produce a gas turbine–powered coupe, which entered the 1963 24 Hours of Le Mans, driven by Graham Hill and Richie Ginther. It averaged 107.8 miles/hour (173 km/hour) and had a top speed of 142 miles/hour (229 km/hour).

General Motors introduced the first commercial gas turbine–powered hybrid vehicle, a limited-production run of the EV-1. A Williams International 40-kW turbine drove an alternator that powered the battery-electric power train. The turbine design included a recuperator.

Gas turbines do offer a high-powered engine in a very small and light package, but there remain unchanged three main reasons why small turbines have not succeeded in an automotive application. First, small turbines are fundamentally less fuel efficient than small piston engines. Second, this problem is exacerbated by the requirement for automotive engines to run efficiently at idle and low throttle openings; turbines are notably inefficient at this. Finally, turbines have historically been more expensive to produce than piston engines, though this is partly because piston engines have been mass-produced in huge quantities for decades, while small turbines are rarities.

A key advantage of jets and turboprops for airplane propulsion—their superior performance at high altitude compared with piston engines, particularly naturally aspirated ones—is irrelevant in automobile applications. Their power-to-weight advantage is far less important. Their use in hybrids reduces the second problem. Capstone currently lists on its Web site a version of its turbines designed for installation in hybrid vehicles.

Their use in military tanks has been more successful. As well as their production use in the T-80 and Abrams, in the 1950s an FV214 Conqueror tank Heavy Tank was experimentally fitted with a Parsons 650-horsepower gas turbine.

A production gas turbine motorcycle first appeared in the MTT Turbine SUPER-BIKE in 2000. This high-priced machine is produced in miniscule numbers.

MICROTURBINES

The small turbine/microturbine mainly consists of a single-stage radial compressor and turbine, a recuperator, and foil bearings.

Figure 3 shows a microturbine designed for the Defense Advanced Research Projects Agency by M-Dot Aerospace. Microturbines are also known as:

- Turbo alternators
- Gensets
- MicroTurbine (registered trademark of Capstone Turbine Corporation)
- Turbogenerator (registered trade name of Honeywell Power Systems, Inc.)

The use of microturbines is becoming widespread for distributed power and combined heat and power applications. They range from handheld units producing less than a kilowatt to commercial-sized systems that produce tens or hundreds of kilowatts.

Part of the success of microturbines is due to advances in electronics, which allow unattended operation and interfacing with the commercial power grid. Electronic power-switching technology eliminates the need for the generator to be synchronized with the power grid. This allows, for example, the generator to be integrated with the turbine shaft, and to double as the starter motor.

Microturbine systems (Figures 4, 5, and 6) have many advantages over piston engine generators, such as higher power density (with respect to footprint and weight), extremely low emissions, and a few, or just one, moving part. Those designed with foil bearings and air-cooling operate without oil, coolants, or other hazardous materials. However, piston engine generators are quicker to respond to changes in output power requirements.

They accept most commercial fuels, such as natural gas, propane, diesel, and kerosene. Microturbines are also able to produce renewable energy when fueled with biogas from landfills and sewage treatment plants.

Microturbine designs usually consist of a single-stage radial compressor, a single-stage radial turbine, and a recuperator. Recuperators are difficult to design and manufacture because they operate under high pressure and temperature differentials. Exhaust heat can be used for water heating, drying processes, or absorption

FIGURE 3 *Microturbine designed for the Defense Advanced Research Projects Agency by M-Dot Aerospace. (Source: DARPA website.)*

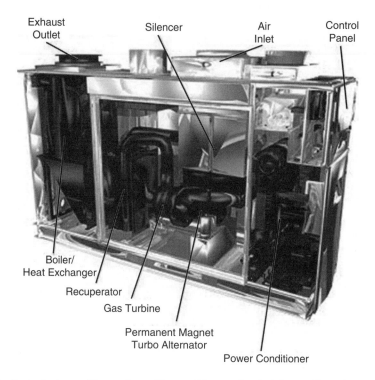

FIGURE 4 *Typical microturbine package. (Source: Thermochemical Power Group.)*

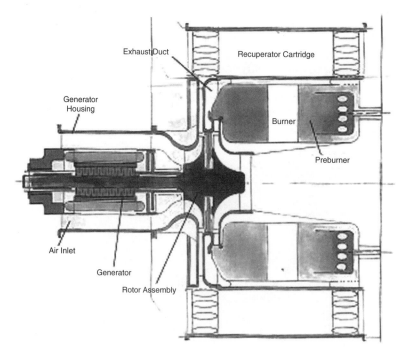

FIGURE 5 *Scheme of microturbine engine. (Source: Thermochemical Power Group.)*

FIGURE 6 *Typical microturbine components. (Source: Thermochemical Power Group.)*

chillers, which create cold for air conditioning from heat energy instead of electric energy.

Typical microturbine efficiencies are 25% to 40%. When in a combined heat and power cogeneration system, efficiencies of greater than 80% are commonly achieved.

Microturbines used with a fuel cell can also reach efficiencies of 80%. A microturbine and a fuel cell may also be termed a *hybrid*. Hybrids can provide power for mobile (automobile) and stationary (power plant on a fixed foundation) applications. For several application case studies, see Part 2 of this book.

ABBREVIATIONS AND ACRONYMS

Most abbreviations and acronyms are explained where they occur in the text. However, there are some other commonly used ones that are listed here:

BAT: best available technology
CHP: combined heat and power
DG: distributed generation
O&M: overhaul and maintenance
OEM: original equipment manufacturer
kW: kilowatt
kWh: kilowatt-hour
MMBTU: million(s) of BTU (British Thermal units)
ROI: return on investment

Part 1

Basic Microturbines

Chapter 1

Distributed Generation and Microturbines

There are various definitions for the term *distributed generation*. For this publication, distributed generation is defined as the use of modular-type electricity generators (such as wind turbines, microturbines, or photovoltaic cells), used independently or as part of a larger system, by utility customers, utilities, and merchant or independent producers whose customers are utilities or utility companies.

The purpose of these generators may be solely to provide electric power or to provide heat and power in combined heat and power (CHP) applications. A prime characteristic of distributed generation is that it does not rely on large transmission lines or a grid as much as conventional power generation does.

Frequently, there may be a tie-in to the local grid for "backup" or "peaking" periods by the end user. The tie-in also allows the small producer to sell power back to the grid, when that producer makes power in excess of its requirements. The primary advantage of distributed generation is the much smaller size of a distributed facility versus a conventional power plant, and therefore its affordability.

Examples of distributed generation include the following:

- A small remote process plant operates using a microturbine solely for its power needs because there is no nearby grid.
- The residents of a Danish village form a cooperative to buy and operate a wind turbine to provide their power needs and have a grid tie-in for when more power is required.
- A Swedish town burns biomass fuel at its local CHP facility that supplies most of its heating and power needs; backup power may be a bank of diesel-fueled generator sets. Note that for CHP applications, the positioning of the buildings supplied with heat is crucial for heat economy: the buildings are normally clustered closely around the primary heat source.
- A restaurant in California uses photovoltaic units on its roof to supply its primary power needs, but maintains a tie-in to the local grid.

Figure 1-1 depicts distributed generation technologies and their applications.

The performance of various distributed generation methods differs in accordance with their best available technology. Figure 1-2 illustrates the market and performance of these methods as of 2000.

Distributed generation is not always the most economic option for power generation, depending on a variety of factors including the logistics of the end-user application in use and the current state of the application's best available technology. Figure 1-3 depicts utility costs in a vertically integrated utility for various production methods in a new production facility as of 2000.

Other factors, such as global conditions at the time (e.g., fuel costs, war or political upheaval in progress) will also change these figures considerably. Additionally, specific

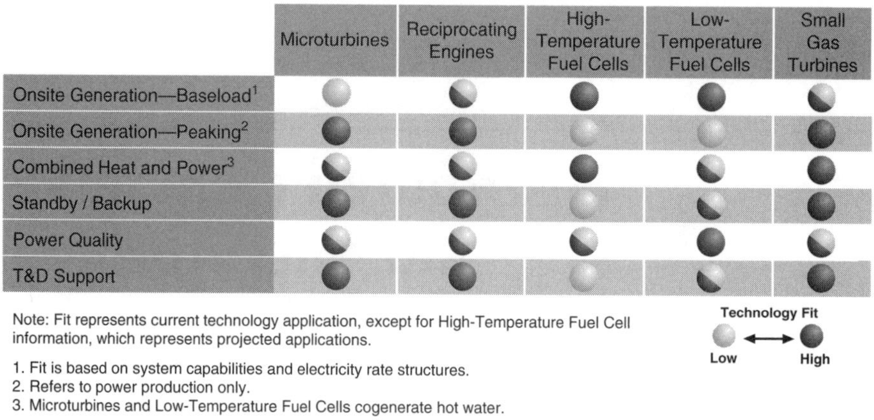

	Microturbines	Reciprocating Engines	High-Temperature Fuel Cells	Low-Temperature Fuel Cells	Small Gas Turbines
Onsite Generation—Baseload[1]					
Onsite Generation—Peaking[2]					
Combined Heat and Power[3]					
Standby / Backup					
Power Quality					
T&D Support					

Note: Fit represents current technology application, except for High-Temperature Fuel Cell information, which represents projected applications.

1. Fit is based on system capabilities and electricity rate structures.
2. Refers to power production only.
3. Microturbines and Low-Temperature Fuel Cells cogenerate hot water.

Technology Fit

Low ◐←→● High

FIGURE 1-1   *Distributed generation technologies and their applications. T&D = transmission and distribution. (Source: Arthur D. Little. 1999.* Distributed energy: understanding the economics. *Arthur D. Little, Inc. Cambridge, MA.)*

	Residential	Commercial	Industrial	Grid-Distributed	Remote / Off-Grid Distributed	Typical Unit Size Range (installation size can be larger)	1999 Installed Capital Cost ($/kW)	Efficency (%)	Commercial Availability
Microturbines[1]		1	1	1	2	25–300 kW	750–900	28–33	1999
Reciprocating Engines	1	1	1	1	1	5 kW–20 MW	400–600[2]	28–37	NOW
High-Temperature Fuel Cells		1	1	1	2	100 kW–1 MW	A[3]	45–55	2005
Low-Temperature Fuel Cells	1	1	2	1	1	2–250 kW	2,000–3,000	30–40	NOW[5]
Small Gas Turbines		1	1	2		500 kW–20 MW	650	25–40[4]	NOW

1. Recuperated microturbine
2. Large, gas-fired reciprocating engine
3. Not available
4. Forty percent efficiency achieved with advanced turbine cycle
5. PAFC only; PEM available in 2000

1 Primary Target Market

2 Secondary Target Market

FIGURE 1-2   *Distributed generation technologies—market and performance. PAFC = phosphoric acid fuel cells; PEM = proton exchange membrane. (Source: Arthur D. Little. 1999.* Distributed energy: understanding the economics. *Arthur D. Little, Inc. Cambridge, MA.)*

system features and characteristics, as well load type (e.g., peaking, base) will alter figures in any reference period. The assumptions made by the data source in constructing Figure 1-3 are described in Table 1-1.

In Figure 1-3, *central plant* (x axis) refers to the traditional power plant model with generating and transmitting power from a central location. *Distributed generation* (x axis) refers to small power generation installed in the distribution system on the utility side of the meter.

Constraint(s) with reference to utilities means that the utility does not have enough capacity to meet the demands placed on it, so investment is required for it to do so. Constraints may occur with respect to generation, transmission, and/or distribution (see Figure 1-3).

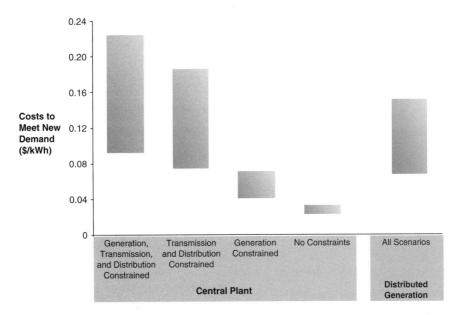

FIGURE 1-3 *Range of utility cost to meet new-demand vertically integrated utility. (Source: Arthur D. Little. 1999.* Distributed energy: understanding the economics. *Arthur D. Little, Inc. Cambridge, MA.)*

TABLE 1-1 Assumptions for Constructing Figure 1-3

Option	Scenarios	Cost Assumptions
1. *Central Plant*	A. *Generation, Transmission, and Distribution*	• Generation—Includes capital and operating costs for a combined-cycle gas turbine (20–60% capacity factor) *Constrained* • Transmission—Includes capital construction costs for new construction and operating costs; Transmission construction costs based on annual capital expenditures and transmission additions for U.S. Investor-Owned Utilities (IOUs) (20–60% capacity factor) *Source: Statistical Yearbook of the Electric Utility Industry 1997, Edison Electric Institute* • Distribution—Includes capital construction costs for new distribution and operating costs; Distribution construction costs based on annual capital expenditures and transmission additions for U.S. IOUs: Operating costs based on annual operation and maintenance expenses and energy sales for IOUs: Also includes secondary distribution system costs for substation upgrades and new construction; Secondary distribution system costs based on interviews and quotes from transformer and substation equipment vendors (20–60% capacity factor) *Source: Statistical Yearbook of the Bedric Utility Industry 1997, Edison Electric Institute*

(Continued)

TABLE 1-1   Assumptions for Constructing Figure 1-3—Cont'd

Option	Scenarios	Cost Assumptions
	B. *Transmission and Distribution Constrained*	• Generation—based on marginal costs of electricity in ERCOT, California, and NEPOOL and market prices from PJM, ISO New England, and California Power Exchange (20–60% capacity factor) • Transmission—Same as 1A • Distribution—Same as 1A • Generation—Same as 1A
	C. *Generation Constrained*	• Transmission and Distribution—Includes transmission and distribution and operating costs based on annual operation and maintenance expenses and energy sales for IOUs *Source: Statistical Yearbook of the Electric Utility Industry 1997, Edison Electric Institute*
	D. *No Constraints*	• Generation—Same as 1B • Transmission and Distribution—Same as 1C
2. *Distributed Generation*	*All*	• Generation—Based on the capital and operating costs for a large natural gas reciprocating engine (20–60% capacity factor) • Distribution—Includes capital costs for the secondary distribution system—substation upgrades and new construction; Secondary distribution system costs based on interviews and quotes from transformer and substation equipment vendors (20–60% capacity factor)

ERCOT, CA = Energy Reliability Council of California; NEPOOL = New England power pool; PJM = Pennsylvania, New Jersey Maryland.
(Source: Arthur D. Little. 1999. *Distributed energy: understanding the economics.* Arthur D. Little, Inc. Cambridge, MA.)

TABLE 1-2   Typical Grid Side Benefits

	$/kW-yr	¢/kWxh (at 60% Capacity Factor)
Avoided Increases in System Capacity	55	1.0
Reduced T&D Losses	50	0.9
T&D Upgrade Deferral	30	0.6
VAR Support	35	0.7
Total EDC Benefits	***170***	***3.2***

VAR = volt-ampere reactive; EDC = electrical distribution companies.
(Source: Arthur D. Little. 1999. *Distributed energy: understanding the economics.* Arthur D. Little, Inc. Cambridge, MA.)

Vertically integrated with respect to power companies means that they own generation, transmission, and distribution assets (see Figure 1-3).

The advantages of distributed generation include the following:

- Power reliability is increased because distributed generation can reduce or avoid outages for specific end users or end-user groups.
- Much of the need for new transmission line system (and associated distribution hardware) construction, is avoided. This then avoids cutting down tracts of virgin rain forest and dealing with communities who object to major alterations in their ecosystem.
- Reserve margins are reduced by lowering overall demand levels for grid power.
- Peak demand power values on the main grid are reduced.
- Power reliability is increased by reducing or avoiding outages in areas served by the main grid.

Table 1-2 attempts to illustrate some of these advantages; however, it should be noted that site, utility, and application factors can radically change these values at any reference time.

Chapter 2

Design and Components
of Microturbines

Microturbines are very small gas turbines (30 to 400 kilowatts [kW]) that usually have an internal heat-recovery heat exchanger (called a *recuperator*) to improve electric efficiency. In typical microturbines, the cycle is similar to that of a conventional gas turbine. It consists of the following processes:

- Inlet air is compressed in a radial (centrifugal) compressor, then
- Preheated in the recuperator using heat from the turbine exhaust.
- Heated air from the recuperator is mixed with fuel in the combustor and burned.

The hot combustion gas is then expanded in one or more turbine sections, which produces rotating mechanical power to drive the compressor and the electric generator. The recuperator efficiency is the key to whether a particular microturbine is economically viable.

By comparison, in a conventional gas turbine,[†] the gas flow path is as follows: compressed air from the compressor (more air mass can be "introduced" by intercooling) is burned with fuel. Gaseous products expand through the turbine section (where more power can be extracted by reheating the gaseous products). Exhaust gases can provide waste heat recovery or cogeneration potential, as those gases may produce steam to drive a steam turbine, may be led into a greenhouse system, or may optimize thermal efficiency by some other means. Without waste heat recovery or cogeneration of some sort, a gas turbine is said to operate in "simple cycle" mode. With the addition of a boiler (to get steam from waste heat recovery) and a steam turbine, the gas turbine package is said to operate as a "combined cycle."

However, most microturbines, to be financially viable, have a recuperator (to recover waste heat). This is not quite a simple cycle, but the microturbine can be said to operate "solo" (see Chapter 9) in power-only applications.

Frequently, microturbines are used to extract heat as a product. This then would be called combined heat and power (CHP) applications (see Chapter 10).

In single-shaft microturbines,[*] a single expansion turbine turns both the compressor and the generator. Two-shaft models use one turbine to drive the compressor and a second turbine to drive the generator, with exhaust from the compressor turbine powering the generator turbine. The power turbine's exhaust is then used in the recuperator to preheat the air from the compressor.

Single-shaft models are designed to operate at high speeds (some in excess of 100,000 revolutions per minute [RPM]) and generate electric power as high-frequency alternating current (AC). The generator output is rectified to direct current (DC) and

---

[*] Adapted from: Gas Research Institute and the National Renewable Energy Laboratory. 2003. *Gas-fired distributed energy resource technology characterizations*. U.S. Department of Energy. Oak Ridge, TN.

[†] Adapted from: Soares, C. 2007. *Gas turbines: a handbook of air, land, and sea applications*. Butterworth-Heinemann. Boston, MA.

then inverted to 60 hertz (Hz) AC for commercial use in the United States, or 50 Hz for use in countries with a 50 Hz supply.

Two-shaft microturbines have a turbine-driven compressor on one shaft and a separated "free" power turbine on a second shaft to power the generator. (In conventional gas turbines, this arrangement is generally used in mechanical drive service, as the free power turbine, generally via a gear box, then runs other turbomachinery, such as compressors or pumps.) With the expansion pressure ratio split between two turbines, the power turbine on a two-shaft machine can be designed to run at lower speed with high efficiency. The power turbine is connected to a conventional 60-Hz AC generator through a low-cost, single-stage gearbox. Some manufacturers offer units producing 50 Hz AC for use in countries where 50 Hz is standard, such as in Europe and Asia. To economize on spares, other original equipment manufacturers (OEMs) have turbine packages for both voltages: they supply a gearbox for one power supply frequency and take the power loss (in the order of 2%) that the gearbox requires.

THERMODYNAMIC HEAT CYCLE

In principle, microturbines and larger gas turbines operate on the same thermodynamic heat cycle, the Brayton cycle. In this cycle, atmospheric air is compressed, heated at constant pressure, and then expanded, with the excess power produced by the expander (also called the *turbine*) consumed by the compressor used to generate electricity.

The power produced by an expansion turbine and consumed by a compressor is proportional to the absolute temperature of the gas passing through those devices. Higher expander inlet temperature and pressure ratios result in higher efficiency and specific power. Higher pressure ratios increase efficiency and specific power until an optimum pressure ratio is achieved, beyond which efficiency and specific power decrease. The optimum pressure ratio is considerably lower when a recuperator is used.

Consequently, for good power and efficiency, it is advantageous to operate the expansion turbine at the highest practical inlet temperature consistent with economic turbine blade materials and to operate the compressor with inlet air at the lowest temperature possible. The general trend in gas turbine advancement has been toward a combination of higher temperatures and pressures. However, microturbine inlet temperatures are generally limited to 1750°F or below to enable the use of relatively inexpensive materials for the turbine wheel and recuperator. For recuperated turbines, the optimum pressure ratio for best efficiency is usually less than 4:1.

See Table 2-1 for a summary of a microturbine's advantages and design characteristics.

MICROTURBINE PACKAGE

The basic components of a microturbine are the compressor, turbine, generator, and recuperator (Figure 2-1). The heart of the microturbine is the compressor-turbine package, which is most commonly mounted on a single shaft along with the electric generator. The single shaft is supported by two (or more) high-speed bearings.

Because single-shaft turbines have only one moving shaft, they have the potential for lower maintenance and higher reliability than turbines with two or more shafts.

There are also two-shaft versions of the microturbine, in which the turbine on the first shaft only drives the compressor while a second power turbine on a second shaft drives a gearbox and conventional electric generator producing 60 or 50 Hz of power. The two-shaft design has more moving parts but does not require power electronics to convert high-frequency AC power output to usable power.

TABLE 2-1 Advantages and Design Characteristics of Microturbines

Thermal output	Microturbines produce waste heat in the exhaust gas at temperatures in the 400°F-to-650°F range, suitable for supplying a variety of building and light industrial thermal needs.
Fuel flexibility	Microturbines can operate using a number of different fuels: natural gas, sour gases (high sulfur, low BTU content), landfill gas, digester gas, and liquid fuels (e.g., gasoline, kerosene, and diesel fuel/heating oil).
Reliability and life	Design life is estimated to be in the 40,000- to 80,000-hour range. Although units have demonstrated reliability, they have not been in commercial service long enough to provide definitive life data.
Size range	Microturbines available or under development are sized from 30 to 400 kW.
Emissions	Sophisticated combustion schemes, relatively low turbine inlet temperatures, and low (lean) fuel-to-air ratios result in NOx emissions of less than 10 parts per million and inherently low CO and unburned hydrocarbon emissions, especially when running on natural gas.
Modularity	Units may be connected in parallel to serve larger loads and to provide power reliability.
Part-load operation	Because microturbines reduce power output by reducing mass flow and combustion temperature, efficiency at part load is below that of full-power efficiency.

Moderate- to large-sized gas turbines use multistage axial flow compressors and turbines, in which the gas flows parallel to the axis of the shaft and then is compressed and expanded in multiple stages. Most current microturbines are based on single-stage radial flow compressors and either single- or double-stage turbines. Radial-flow turbomachinery can handle the very small volumetric flows of air and combustion products with higher component efficiency and with simpler construction than axial-flow components.

With axial-flow turbomachinery, blade height would be too small to be practical or efficient. Although large-sized axial-flow turbines and compressors are typically more efficient than radial-flow components, in the size range of microturbines (0.5 to 5 pounds per second of air/gas flow) radial-flow components offer lower surface and end

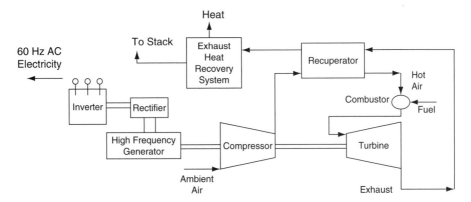

FIGURE 2-1 *Microturbine-based combined heat and power system (single-shaft design). (Source: Gas Research Institute and the National Renewable Energy Laboratory. 2003.* Gas-fired distributed energy resource technology characterizations. *U.S. Departmenet of Energy. Oak Ridge, TN.)*

wall losses and provide higher efficiency. Radial-flow components generally cost less to manufacture.

In microturbines, the turbo-compressor shaft turns at about 96,000 RPM in 30-kW machines and about 70,000 RPM in 100-kW machines. Although rotational speed generally increases as physical size decreases, there is no single rotational speed-power size rule. The specific turbine and compressor design characteristics for a given power output influence the physical size of components and consequently rotational speed.

As aerodynamic equations would indicate, for any turbine design, as the power rating and the compressor and turbine diameters decrease, the shaft speed required to maintain turbine operability increases—hence the very high shaft speed of the very small microturbines.

Radial-flow turbine-driven compressors are quite similar in terms of design and volumetric flow to turbochargers used in automobiles, trucks, and other small reciprocating engines. Superchargers and turbochargers have been used for almost 80 years to increase the power of reciprocating engines by providing compressed inlet air to the engine.

Today's worldwide market for small automobile and truck turbochargers is around two million units per year and is characterized by relatively low prices to the vehicle engine OEMs. Very small gas turbines (of the size and power rating of microturbines) have been used extensively in auxiliary power systems and auxiliary power units on airplanes. Decades of experience with these applications provide the basis for the engineering and manufacturing technology of emerging microturbine models.

RECUPERATORS

Recuperators are air-to-gas heat exchangers that use the hot turbine exhaust gas (typically around 1200°F) to preheat the compressed air (typically around 300°F to 400°F) before the compressed air goes into the combustor, thereby reducing the fuel needed to heat the compressed air to the design turbine inlet temperature. Microturbines require a recuperator to achieve the efficiency levels needed to be competitive in continuous duty service.

Note in Figure 2-2, the feature of high effectiveness (up to approximately 90%). *Effectiveness* is the technical term in the heat-exchanger industry for the ratio of the actual heat transferred to the maximum achievable heat transferred. A recuperator increases the electric efficiency of a microturbine from about 14% to about 26% for a typical turbine depending on component details.

However, incorporation of a recuperator results in pressure losses through the recuperator itself and in the ducting that connects it to other components. The pressure loss essentially reduces the pressure ratio available to the turbine, affecting both power output and system efficiency. Typically, these pressure losses result in 10% to 15% less power being produced by the microturbine and a corresponding loss of a few points in efficiency compared with an ideal recuperator with a zero pressure drop.

Although microturbine performance could be enhanced by using recuperators with higher effectiveness and lower pressure drop, such recuperators would be large and expensive.

Effective optimization of recuperator design and integration must balance performance and cost. These must be based on the economic factors of the application (the relative values of electric and thermal energy) rather than from the consideration of the recuperator component alone.

Recuperator durability is also a design and economic issue for microturbine developers. Recuperators previously in use on industrial gas turbines developed leaks due to differential thermal expansion accompanying thermal transients. However, microturbine recuperators have proven quite durable in testing and commercial applications to date. This improvement in durability has resulted from the use of higher-strength alloys and higher-quality welding, along with improved engineering design to avoid the

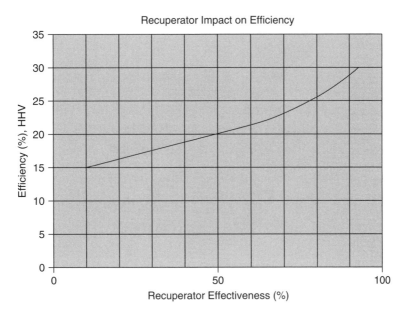

FIGURE 2-2 *Microturbine efficiency as a function of recuperator effectiveness. HHV = higher heating value. (Source: Gas Research Institute and the National Renewable Energy Laboratory. 2003. Gas-fired distributed energy resource technology characterizations. U.S. Department of Energy. Oak Ridge, TN. Taken from: Energy and Environmental Analysis, Inc., estimates, based on turbine pressure ratio of 3.2.)*

internal differential expansion that causes stresses resulting in cracks and leakage. Such practical improvements can come at an appreciable cost, again affecting the economic attractiveness of the microturbine.

Recuperator Construction Details

The following extract describes one OEM's manufacturing strategy in dealing with common recuperator design issues. Further testing conducted by this OEM, to mitigate the effects of environmental degradation in recuperators, is briefly mentioned here and discussed further in Chapter 6, Microturbine Performance Optimization and Testing.

The primary surface recuperator, or PSR (Solar Turbines Inc. design[*]), allows for heat transfer between the exhaust and compressor discharge gas streams in a gas turbine (Figure 2-3). The PSR design has an inherently large heat transfer surface area-to-volume ratio that allows for high efficiency in a relatively compact package. Design innovations reduce the effects of temperature gradients present across the recuperator, thus limiting thermal deformation and thermal fatigue. The PSR design offers relatively low maintenance and extended service life compared with other common types of heat transfer devices.

Pairs of these sheets are welded to header bars located around the perimeter to form an air cell (Figure 2-4), the basic building block of the PSR. Each air cell is pressure-checked

---

[*] Source: Rakowski, J. M., Stinner, C. P. (TI Allegheny Ludlum), Lipschtz, M., and Montague, J. P. (Solar Turbines Inc.). 2004. *The use and performance of oxidation and creep-resistant stainless steels in an exhaust gas primary surface recuperator application.* International Gas Turbine Institute (IGTI) GT 2004-53917.

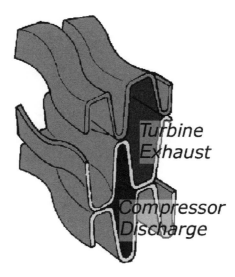

FIGURE 2-3   *Folded primary surface (schematic). (Source: Rakowsky, J. M., Stinner, C. P., Lipshutz, M., and Montague, J. P. 2004. The use and performance of oxidation and creep-resistant stainless steels in an exhaust gas primary surface recuperator application.* Proceedings of ASME Turbo Expo 2004. *American Society of Mechanical Engineers. Vienna, Austria.)*

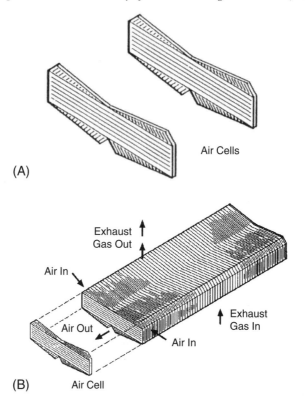

FIGURE 2-4   *(A) Air cells, (B) Core assembly (schematics). (Source: Rakowsky, J. M., Stinner, C. P., Lipshutz, M., and Montague, J. P. 2004. The use and performance of oxidation and creep-resistant stainless steels in an exhaust gas primary surface recuperator application.* Proceedings of ASME Turbo Expo 2004. *American Society of Mechanical Engineers. Vienna, Austria.)*

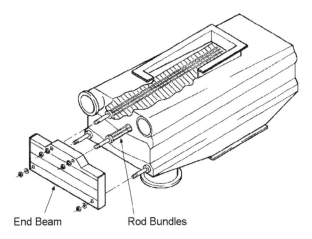

End Beam   Rod Bundles

FIGURE 2-5 *Thermally balanced restraint system. (Source: Rakowsky, J. M., Stinner, C. P., Lipshutz, M., and Montague, J. P. 2004. The use and performance of oxidation and creep-resistant stainless steels in an exhaust gas primary surface recuperator application.* Proceedings of ASME Turbo Expo 2004. *American Society of Mechanical Engineers. Vienna, Austria.)*

before it is welded into the recuperator core assembly (Figure 2-5). There are no internal welds or joints within the air cell.

The header bars in the individual air cells are designed to resist internal pressure without added support, but the primary sheets require full support to prevent "ballooning" under internal pressure. A thermally balanced restraint system is added to the core to contain the pressure forces (see Figure 2-5). Clamping the recuperator core between two rigid plates (or *end beams*) that cover the ends of the core provides the necessary support.

Preloaded tie rods connect the two end beams and provide the clamping force. The tie rods have thermal expansion characteristics that closely match those of the core so that constant support is provided during thermal transients. To improve thermal response, the tie rods are placed in the gas stream. Each rod assembly consists of a bundle of small-diameter rods across which exhaust gas can flow.

This external restraint philosophy has two major advantages. First, it eliminates the need for internal structures within air cells and the recuperator core, which would otherwise be required to hold the core together. Second, the external restraint system relieves the recuperator air and exhaust ducting of any requirement to provide structural support. This sheet metal ducting is added to the recuperator core as required to direct the air and exhaust gases through the recuperator assembly. Use of thin duct walls enhances the flexibility of the core and its ability to accommodate the natural curvature imposed on the core by its internal temperature differential without inducing high stresses.

Solar Turbines Inc. PSRs have accumulated thousands of hours of field operation, successfully validating the recuperator design. However, degradation of Type 347 recuperator material has been observed in post-service inspections. The primary damage mechanisms for PSR foil, welds, and ducting are creep and oxidation. Creep of Type 347 thin foil within individual air cells during operation has been shown to alter the shape of the corrugated pattern (Figure 2-6). Severe creep deformation can restrict gas flow, increase turbine back-pressure, and decrease overall efficiency.

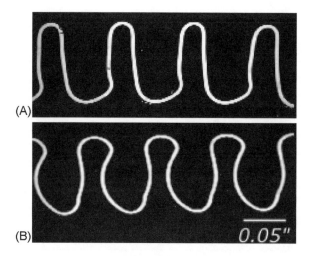

FIGURE 2-6   *Corrugated Type 347 PSR foil before (A) and after (B) service. (Source: Rakowsky, J. M., Stinner, C. P., Lipshutz, M., and Montague, J. P. 2004. The use and performance of oxidation and creep-resistant stainless steels in an exhaust gas primary surface recuperator application. Proceedings of ASME Turbo Expo 2004. American Society of Mechanical Engineers. Vienna, Austria.)*

BEARINGS[*]

Microturbine shafts may be supported on either oil-lubricated or air bearings. *Oil-lubricated bearings* are mechanical bearings and come in three main forms: high-speed metal ball or roller, floating sleeve, and ceramic surface. The last of these typically offers the most attractive benefits in terms of life, operating temperature, and lubricant flow. Although oil-lubricated bearings are a well-established technology, they require an oil pump, oil-filtering system, and oil cooling, which add to microturbine cost and maintenance.

Also, the exhaust from machines featuring oil-lubricated bearings may not be usable for direct-process heating in cogeneration configurations due to the potential for contamination of the exhaust. Since the oil never comes in direct contact with hot combustion products, as is the case in small reciprocating engines, the expected relatively low oil contamination and replacement requirements of such a lubrication system are more typical of ship propulsion diesel systems (which have separate bearings and cylinder lubrication systems) and automotive transmissions than of automotive engine lubrication.

Air bearings have been in service on airplane auxiliary power systems and cabin-cooling systems for many years. They allow the high-speed turbine to be aerodynamically supported on a thin layer of air, so friction is low. No oil, oil pump, or cooling system is required. Air bearings offer simplicity of operation without the cost, reliability concerns, maintenance requirements, or power drain of an oil supply and filtering system.

Concern does exist for the reliability of air bearings under numerous and repeated starts due to metal-on-metal friction during start-up and shutdown. Reliability depends to a large extent on individual manufacturers' quality control methodology and on design engineering—and it will only be proven after significant experience with substantial numbers of units with long operating times and many on-off cycles.

Field experience with microturbines is growing rapidly, and one manufacturer has successfully demonstrated more than 10,000 start-load-stop cycles on numerous commercial service microturbines equipped with air bearings.

---

[*] Adapted from: Gas Research Institute and the National Renewable Energy Laboratory. 2003. *Gas-fired distributed energy resource technology characterizations*. U.S. Department of Energy. Oak Ridge, TN.

GENERATOR

The microturbine produces electric power either via a high-speed generator directly driven by the single turbo-compressor shaft or with a separate power turbine driving a gearbox and conventional 3600-RPM generator. The high-speed generator of the single-shaft design makes use of a permanent magnet (typically samarium cobalt) alternator and requires that the high-frequency AC output (about 1600 Hz for a 30-kW machine) be converted to 60 Hz for general use.

This power conditioning involves rectifying the high-frequency AC to DC, and then inverting the DC to 60 Hz AC. Frequency conversion comes with a small efficiency penalty (on the order of 5%). To start the single-shaft design, the generator serves as a motor turning the turbo-compressor shaft until sufficient speed (RPM) is reached to allow the compressor to supply enough air for self-sustaining combustion.

If the system is operating independent of the grid (i.e., black-start capability), a power storage unit (typically a battery uninterruptable power supply [UPS]) is used to power the generator for start-up. Two-shaft designs use a separate starting system.

Single-shaft microturbines have digital power controllers to convert the high-frequency AC power produced by the generator into commercially viable electricity. The high-frequency AC is rectified to DC, inverted back to 60 or 50 Hz AC, and then filtered to reduce harmonic distortion.

Power electronics are a critical component in the single-shaft microturbine design and represent significant design challenges, specifically in matching turbine output to the required load. Power electronics are generally designed to handle seven times the nominal voltage to allow for transients and voltage spikes. Most microturbine power electronics are designed for generating three-phase electricity.

Electronic components control all of the engine/generator operating and start-up functions. Microturbines are generally equipped with controls that allow the unit to be operated either in parallel with or independent of the grid, and they incorporate many of the grid and system protection features required for interconnection. The controls usually also allow for remote monitoring and operation.

INLET AIR COOLING

The decreased power and efficiency of microturbines at high ambient temperatures indicate that microturbine electric-power output is at its lowest when power is typically in greatest demand and most valued (e.g., summer peak periods). The use of inlet air cooling can mitigate the decreases in power and efficiency resulting from high ambient air temperatures.

Although inlet air cooling is not a feature on today's microturbines, cooling techniques now entering the market on large gas turbines can be expected to work their way to progressively smaller equipment sizes. At some future date, these will be used with microturbines, if economically viable.

Evaporative cooling, a relatively low capital/cost technique, is the most likely to be applied to microturbines. A fine spray of water is injected directly into the inlet air stream, and evaporation of the water cools the air.

Since cooling is limited to the wet-bulb air temperature, evaporative cooling is most effective when the wet-bulb temperature is appreciably below the dry-bulb (i.e., ordinary) temperature. In most locales with high daytime dry-bulb temperatures, the wet-bulb temperature is often 20°F lower. This affords an opportunity for substantial evaporative cooling and increased microturbine output. However, evaporative cooling can consume large quantities of high-quality water, making it a difficult and costly application in arid climates.

Refrigeration cooling in microturbines is also technically feasible. In refrigeration cooling, a compression-driven or thermally activated (i.e., absorption) refrigeration device cools the inlet air through a heat exchanger. The heat exchanger in the inlet air stream causes an additional pressure drop in the air entering the compressor, thereby slightly lowering cycle power and efficiency. As the inlet air is now substantially cooler than the ambient air, there is a significant net gain in power and efficiency.

Electric motor compression refrigeration entails a substantial parasitic power loss. Thermally activated absorption cooling can use waste heat from the microturbine, reducing the direct parasitic loss. However, the relative complexity and cost of these approaches, in comparison with evaporative cooling, make them less likely to be incorporated into microturbines.

Finally, it is also technically feasible to use thermal-energy storage systems—typically, ice, chilled water, or other low-temperature fluids—to cool inlet air. These systems can substantially reduce parasitic losses from the augmented power capacity. Thermal energy storage is a viable option if on-peak power pricing only occurs a few hours a day. In that case, the shorter time of energy storage discharge and longer time for daily recharging allow for a smaller and less expensive thermal-energy storage system.

FIRING (TURBINE INLET) TEMPERATURE

Microturbines are limited to firing temperatures within the capabilities of available gas turbine alloys. This is in contrast to industrial and utility turbines (25 to 2000 pounds per second of mass flow), which are usually equipped with internal cooling capability to permit operation with firing temperatures well above those of the service limits of alloys used in gas turbines.

Design improvements that enable increased large gas turbine efficiency, via higher firing temperatures, have occurred more through the development and advancement of blade and vane internal cooling technology than through the improvement of the high temperature capabilities of gas turbine alloys.

However, the nature of the three-dimensional shape of radial inflow turbines used in microturbines and their small physical size have not yet allowed the development of cost-effective manufacturing methods that can produce internal cooling. This is the primary driver for ceramic technology development for future microturbines. High firing temperature also affects recuperator materials. As the turbine inlet temperature increases, so will turbine outlet temperature, thereby increasing thermal stress and oxidation rates in the recuperators.

FUEL GAS COMPRESSORS

Microturbines require gaseous fuel to be supplied in the 64- to 90-pounds per square inch gauge range (or above). Most local gas-distribution systems operate well below this range. Rotary vane, scroll, and screw compressors have been used to boost gas pressure at the site to the pressure needed by the microturbine.

COMBINED HEAT AND POWER OPERATION

In combined heat and power (CHP) operation, a second heat-recovery heat exchanger—the exhaust gas heat exchanger—can be used to transfer remaining energy from the microturbine exhaust to a hot water system. Exhaust heat can be used for a number of different applications, including process or space heating, heating potable water, driving absorption chillers, or regenerating desiccant dehumidification equipment.

Some microturbine-based CHP applications do not use recuperators, and some have the ability to bypass their recuperator to adjust their thermal-to-electric ratio. The temperature of the exhaust from these microturbines is much higher (up to 1200°F) and thus more heat is available for recovery. A microturbine-based CHP system is illustrated in Figure 2-1.

Chapter 3

Microturbine Application and Performance

Microturbines are well suited for a variety of distributed generation applications due to their:

- Flexibility in connection methods,
- Ability to be arranged in parallel to serve larger loads,
- Ability to provide reliable power, and
- Low emissions profile.

Potential microturbine applications include premium and remote power, as well as grid support. Combined heat and power (CHP) applications produce power and use exhaust gas heat to:

- Produce domestic hot water,
- Heat building space,
- Drive absorption cooling or desiccant dehumidification equipment, and/ or
- Supply other thermal energy needs in a building or industrial process.

Target customers for microturbine system applications are in light industrial facilities and in financial services, data processing, telecommunications, health care, lodging, retail, offices, schools, and other commercial or institutional buildings.

POWER-ONLY*

Peaking

In certain areas, customers and utilities are using on-site power generation to reduce the cost of peak-load power. Peak shaving is applicable to customers with poor load factor, high electricity demand charges, or both. Typically, peak shaving does not involve heat recovery, although thermal recovery may be cost effective if the peak period is more than 2000 hours per year. Generally, low equipment cost and high reliability are the primary requirements for peaking applications. Emissions may be an issue if the annual number of operating hours is high or the operating facility is located in a nonattainment area. Where peak shaving can be combined with another function, such as standby or emergency power, the economics are considerably enhanced.

There are three peak-shaving strategy alternatives:

- First, customers can independently optimize the use of purchased versus generated power under applicable rate structures. Under this strategy, on-site generation would

---

* Adapted from: Gas Research Institute and the National Renewable Energy Laboratory. 2003. *Gas-fired distributed energy resource technology characterizations*. U.S. Department of Energy. Oak Ridge, TN.

21

operate during the utility-defined, more costly peak periods. This creates an operating strategy that can vary, depending on the tariff, from 900 hours per year to as many as 3500 hours per year.

- Second, some utilities offer coordinated peak-shaving programs in which the utility offers payments to the customer for very limited hours of use on request from the utility. These programs typically require that the system operate between 50 to 400 hours per year.
- Third, for customers who purchase power competitively, there is an opportunity to produce their own power more cheaply during the hours of high peak pricing or to select more competitive, interruptible power-supply contracts.

In the competitive market strategy, the hours of operation probably would be closer to the reduced hours of the coordinated utility program than the independent peak shaving of a published tariff. The optimal technology configuration and the need to integrate the value of standby power with the value of peak shaving differ markedly among these operating strategies.

In general, current market-entry microturbines are too costly to provide economic peaking power in most regions of the United States. However, peaking may develop into a significant microturbine application if the potential for lower capital costs and increased efficiencies can be achieved as microturbine markets develop and the technology matures.

Premium Power

Consumers who require higher levels of reliability or power quality, and are willing to pay a premium for it, often find some form of distributed generation to be advantageous. These consumers are typically less concerned than other electrical supply consumers about the initial prices of power-generating equipment. In the premium power market, the current prices of microturbines may be justified, based on their potential advantages in terms of low emissions, reduced vibration, ease of installation, potential for high availability and reliability, and good power quality. As is often the case in premium power applications, noise, vibration, and emissions can be significant concerns. As trading emissions increasingly becomes a factor in power generation economics, the emissions factor rises in importance.

Remote Power

In locations where power from the local grid is unavailable or extremely expensive to install, microturbines can be a competitive option. Electric service points tend to be far apart in remote locations. Thus, connection to the local grid by installing an electrical distribution system from the grid to small users can be expensive.

The allowable initial costs of connecting to the grid are often higher than installation of a distributed generation system. As with premium power, remote power applications also are generally base-load operations. As a result, on a long-term basis, emissions and fuel-use efficiency become more significant criteria in much of the remote power market.

However, remote power applications also include oil- and gas-production fields, wellheads, coal mines, and landfills, where by-product gases serve as readily available, low-cost fuel. These locations are often far from the grid and, even when served by the grid, may experience costly downtime when electric service is lost due to weather, fire, or animals. The fuel flexibility of microturbines becomes a significant asset in these facilities, as evidenced by the early market success of microturbines in these applications.

Grid Support

A growing number of utilities use diesel or natural gas engines at substations to provide incremental peaking capacity and grid support. Such installations can defer the need for transmission and distribution system upgrades, can provide temporary peaking capacity within constrained areas, or can be used for system power factor correction and voltage support, thereby reducing costs for both customers and the utility system.

Microturbines are suitable for such service as well, particularly if multiple units are used to increase available capacity and site reliability.

COMBINED HEAT AND POWER

CHP systems combine on-site power generation with the use of by-product heat. The economics of microturbine power generation are enhanced by continuous base-load operation and the effective use of the thermal energy contained in the exhaust gas. Heat can generally be recovered in the form of hot water or low-pressure steam (<30 pounds per square inch gauge [psig]), or the hot exhaust can be used directly for applications such as process heating or drying (e.g., grain drying, brick drying, or greenhouses). The exhaust heat can be used to drive thermally activated equipment, such as absorption chillers for cooling or desiccant wheel regeneration for dehumidification.

Commercial/institutional buildings and light industrial facilities whose electrical loads and space heating, hot water, or other thermal needs occur at the same time are usually best suited for microturbine CHP applications. The simplest thermal requirement to supply is hot water.

Primary applications for microturbine CHP in the commercial/institutional sectors are those building types with relatively high and coincident electric and hot water demand, such as colleges and universities, hospitals and nursing homes, and lodging buildings. Office buildings and certain warehousing and mercantile/service applications also may be economic applications for microturbine CHP systems because noise and emissions are often siting and permitting issues.

Primary CHP applications in the light industrial market include food processing, chemicals manufacturing, and plastics-forming plants with hot water or low-pressure steam demands. Heat-activated cooling, refrigeration, and desiccant technologies that are now being developed for use with engine-driven systems will broaden microturbine CHP applications by increasing the thermal energy loads in certain building types such as restaurants, supermarkets, offices, and refrigerated warehouses.

COST AND PERFORMANCE CHARACTERISTICS

This subsection describes the cost and performance of microturbine systems for two primary applications. The first is for systems designed to produce power only. Systems configured for this purpose could be used in a variety of applications, including premium power, peaking, and grid support.

The second configuration discussed in this section is CHP, where additional equipment is added to the basic microturbine to allow recovery and subsequent use of exhaust heat in industrial processes or commercial buildings. CHP applications are the primary market for early-entry microturbine systems because use of the waste heat can significantly enhance project economics.

System Performance

The thermal efficiency of the Brayton cycle is a function of pressure ratio, ambient air temperature, turbine inlet air temperature, efficiency of the compressor and turbine elements, and any performance enhancements such as heat recuperation. The addition of the recuperator makes microturbines more complex than conventional simple-cycle gas turbines.

The recuperator reduces fuel consumption (thereby substantially increasing efficiency) but also introduces additional internal pressure losses that moderately lower efficiency and power. Because the recuperator has four connections—to the compressor discharge, the expansion turbine discharge, the combustor inlet, and the system exhaust—it becomes a challenge to microturbine product designers to integrate recuperation in a manner that minimizes pressure loss, keeps manufacturing cost low, and entails the least compromise of system reliability when operated and thermally cycled for extended periods. Each manufacturer's models have evolved in unique ways to meet these challenges.

Table 3-1 provides cost and performance characteristics for typical emerging microturbine CHP systems. The range of 30 to 100 kilowatts (kW) represents what is currently or soon to be commercially available.

Microturbine package performance characteristics are shown in the top portion of the table and are applicable to both power-only and CHP applications. Total installed-cost estimates are provided for both applications. Heat rates and efficiencies shown were taken from manufacturers' specifications and industry publications. Electrical efficiencies are the net of the parasitic and conversion losses, as indicated in a footnote to Table 3-1.

Available thermal energy is calculated based on manufacturer specifications on turbine exhaust flows and temperatures. CHP thermal recovery estimates are based on producing hot water for process or space-heating applications and a heat recovery unit exhaust temperature of 220°F. Total CHP efficiency is defined as the sum of the net electricity generated plus the useful thermal energy (i.e., hot water) produced for thermal needs, as a fraction of total fuel input energy.

Microturbine systems can have different electrical and CHP efficiencies based on the specific design parameters selected by the manufacturer. Each microturbine manufacturer represented in Table 3-1 uses a unique recuperator configuration, and each has made individual trade-offs between cost and performance.

The cost and performance trade-offs include:

- The extent to which the recuperator effectiveness increases turbine efficiency,
- The extent to which the recuperator pressure drop decreases power, and
- The chosen turbine pressure ratio.

Microturbines typically require 50 to 90 psig fuel supply pressure. Because current microturbines have pressure ratios between 3 and 4 to maximize efficiency with use of a recuperator at modest turbine inlet temperature, the required supply pressure for microturbines is much less than for industrial-size gas turbines, which normally have pressure ratios of 7 to 35. Local distribution system gas pressures usually range from 30 to 130 psig in feeder lines and from 1 to 50 psig in final distribution lines.

Most U.S. businesses that would use 30-, 70-, or 100-kW microturbines receive gas at about 12 to 24 inches of water pressure (0.5 to 1.0 psig). Most building codes prohibit piping natural gas into buildings at more than 2 psig. Thus, microturbines in most commercial locations require a fuel gas booster compressor (GBC) to ensure that fuel pressure is adequate for the gas turbine flow-control and combustion systems.

TABLE 3-1   Microturbine Systems—Typical Performance Parameters (2003)

Cost and Performance Characteristics[1]	System 1	System 2	System 3	System 4
Nominal Electricity Capacity (kW)	30 kW	70 kW	50 kW	100 kW
Microturbine Characteristics				
Net Electrical Capacity (kW)[2]	28	67	76	100
Electrical Efficiency (%), LHV	25.1	28.0	26.9	28.9
Electric Heat Rate (Bm/kWh), HHV[3]	15,071	13,544	14,103	13,127
Electrical Efficiency (%), HHV[4]	22.6	25.2	24.2	26.0
Fuel Input (MMBtu/hr)	0.423	0.91	1.09	1.31
Installed Cost – Power Only (2003 S/kW)	2263	1708	1713	1576
Installed Cost - CHP (2003 S/kW)[5]	2636	1926	1932	1769
O&M Costs (2003 S/kWh)	0.02	0.015	0.013	0.015
Required Fuel Gas Pressure (psig)[6]	55	70	85	90
Required Fuel Gas Pressure w/GBC (psig)[7]	0.2–15	0.2–15	0.2–15	0.3–15
CHP Characteristics				
Exhaust Flow (lbs/sec)	0.68	1.60	1.67	1.76
Turbine Exhaust Temperature (°F)	530	450	500	520
Heat Exchanger Exhaust Temperature (°F)[8]	220	220	220	220
Heat Output (MMBTU/hr)	0.186	0.325	0.412	0.466
Heat Output (kW equivalent)	54	95	121	136
Total CHP Efficiency (%), HHV[9]	67	61	63	62
Heat/Fuel Ratio[10]	0.44	0.36	0.38	0.35
Power/Heat Ratio[11]	0.52	0.70	0.63	0.73
Net Heat Rate (Btu/kWh)[12]	6795	7485	7320	7300

MMBTU = millions of BTU; CHP = combined heat and power; O&M = overhaul and maintenance; kWh = kilowatt hours.

[1]Table data are based on Capstone Model 330-30 kW; IR Energy Systems 70LM-70 kW (two-shaft); Bowman TG80-80 kW; Turbec T100-100 kW. Performance characteristics are based on International Organization for Standards standard ambient temperature of 59°F.

[2]Net of parasitic losses from gas booster compressor (GBC) and conversion losses from power conversion equipment.

[3]All turbine and engine manufacturers quote heat rates in terms of the lower heating value (LHV) of the fuel. On the other hand, the usable energy content of fuels is typically measured on a higher-heating-value (HHV) basis. In addition, electric utilities measure power plant heat rates in terms of HHV. For natural gas, the average heat content of natural gas is 1030 British thermal units per standard cubic foot (BTU/scf) on an HHV basis and 930 BTU/scf on an LHV basis, or about a 10% difference.

[4]Electrical efficiencies are the net of parasitic and conversion losses. Fuel gas compressor needs are based on a 1–pound per square inch inlet supply.

[5]Installed costs are based on a CHP system producing hot water from exhaust heat recovery. The 70-, 80-, and 100-kW systems are being offered with integral hot water recovery built into the equipment. The 30-kW units are currently built as electric (only) generators, and the heat recovery water heater is a separate unit.

[6]Fuel gas pressure required at the combustor. This value determines the GBC requirements at a specific site.

[7]Fuel gas pressure required to the GBC.

[8]Heat recovery is calculated based on hot water production (160°F to 180°F) and heat-recovery unit exhaust temperature of 220°F.

[9]Total CHP efficiency = (net electric generated + net heat produced for thermal needs)/total system fuel input.

[10]Heat/fuel ratio = thermal output (total heat recovered MMBTU/hr)/fuel input (MMBTU/hr).

[11]Power/heat ratio = CHP electrical power output (BTU equivalent)/useful heat output (BTU).

[12]Net heat rate = (total fuel input to the CHP system—fuel normally used to generate the same amount of thermal output as the CHP system output, assuming efficiency of 80%)/CHP electric output (kW).

(Source: Gas Research Institute and the National Renewable Energy Laboratory. 2003. *Gas-fired distributed energy resource technology characterizations*. U.S. Department of Energy. Oak Ridge, TN.)

Energy and Environmental Analysis, Inc. (EEA) estimates of characteristics are representative of "typical" commercially available or soon-to-be-available microturbine systems.

Most microturbine manufacturers offer the equipment package with an optional GBC, if needed, to satisfy the microturbine fuel inlet pressure requirement. The GBC is included in the package cost of all of the representative systems shown in Table 3-1.

This packaging facilitates the purchase and installation of a microturbine because the burden of obtaining and installing the booster compressor is no longer placed on the customer. Also, it is believed to result in higher reliability of the booster through standardized design and volume manufacture.

Booster compressors can add from $50 to $180 per kW to a microturbine power-only and CHP system's total cost. As well as adding to capital cost, booster compressors decrease net power and efficiency, making operating cost slightly higher. Typically, about 5% of the output of a microturbine is needed to power the fuel gas booster drive.

Such power loss results in a penalty on efficiency of about 1.5 percentage points. For installations where the unit is located outdoors, customers might be able to save on cost and operating expense by having the gas utility deliver gas at an adequate pressure that eliminates the need for the fuel GBC.

MICROTURBINE DESIGN CONSIDERATIONS

The addition of a recuperator to a gas turbine opens numerous design parameters to performance-cost trade-offs. In addition to selecting the pressure ratio for high efficiency and best-business opportunity (high power for low price), the recuperator has two performance parameters—effectiveness and pressure drop—that also have to be selected for the combination of efficiency and cost that creates the optimum conditions for the application. More effective recuperation requires greater recuperator surface area, which both increases cost and incurs additional pressure drop. Increased pressure drop reduces the net power produced and consequently increases microturbine cost per kW.

Microturbine performance, in terms of both efficiency and specific power (*specific power* is power produced by the machine per unit of mass flow through the machine), is highly sensitive to small variations in component performance and internal losses. This is because the high-efficiency recuperated cycle processes a much larger amount of air and combustion products flow per kW of net mechanical power delivered than is the case for high-pressure-ratio nonrecuperated gas turbines.

As the net power output is the small difference between two large values (the compressor and expansion turbine work per unit of mass flow), small losses in compressor or turbine efficiency, internal pressure losses, and recuperator effectiveness have large impacts on net system efficiency and power per unit of mass flow. Additionally, variances in manufacturing tolerances can have greater effect on component efficiencies on smaller turbomachinery such as microturbines.

Estimated recuperated microturbine electrical efficiency is shown in Figure 3-1 as a function of microturbine compressor ratio, for an ambient air temperature of 59°F (International Organization for Standards [ISO] standard) and a range of turbine inlet temperatures from 1550°F to 1750°F, corresponding to a range of conservative to optimistic turbine material life behavior. The efficiency shown in Figure 3-1 is defined on the basis of generator output without parasitic or conversion losses considered.

Often this output is at high frequency, so it must be rectified and inverted to provide 60 Hz of alternate current power. The efficiency loss in such frequency conversion (about 5%, which would lower efficiency from 30% to 28.5%) is not included in these charts. Figure 3-2 shows that a broad near-optimum of performance exists in the pressure ratio range of 2.5 to 4.5.

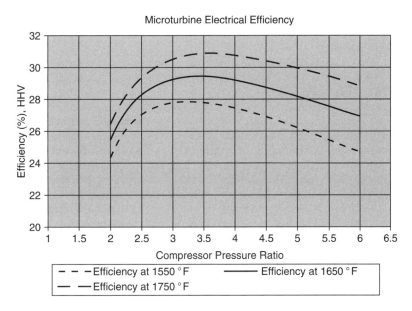

FIGURE 3-1 *Typical microturbine efficiency as a function of compressor pressure ratio and turbine inlet temperature. (Source: Energy and Environmental Analysis, Inc., estimates.)*

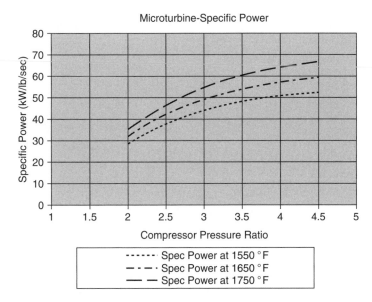

FIGURE 3-2 *Microturbine-specific power as a function of compressor pressure ratio and turbine inlet temperature.*

Figure 3-2 shows microturbine-specific power for the same range of firing temperatures and pressure ratios as in Figure 3-1. Greater specific power can be obtained with higher pressure ratios. Higher pressure ratios are developed by higher rotational speeds, but these speeds are limited by the stress that can be tolerated by practical turbine blade materials. These considerations along with manufacturing cost trade-offs limit compressor pressure ratios to 3:1 to 4:1 in microturbines currently entering the market.

PART-LOAD PERFORMANCE

When less than full power is required from a microturbine, the output is reduced by decreasing rotational speed, which reduces temperature rise and pressure ratio through the compressor and temperature drop through the turbine, and by reducing turbine inlet temperature so that the recuperator inlet temperature does not rise. In addition to reducing power, this change in operating conditions also reduces efficiency.

The efficiency decrease is minimized by the reduction in mass flow (through speed reduction) at the same time as the turbine inlet temperature is reduced. Figure 3-3 shows a typical part-load efficiency curve based on a 30-kW microturbine.

EFFECTS OF AMBIENT CONDITIONS ON PERFORMANCE

The ambient conditions at the inlet of the microturbine affect both the power output and efficiency. At inlet air temperatures above 59°F, both the power and efficiency decrease. The power decreases due to the decreased air density with increasing temperature, and the efficiency decreases because the compressor requires more power to compress higher temperature air.

Conversely, the power and efficiency increase when the inlet air temperature is below 59°F. Figure 3-4 shows typical variation in power and efficiency for a micro-turbine as a function of ambient temperature, relative to the reference ISO condition of

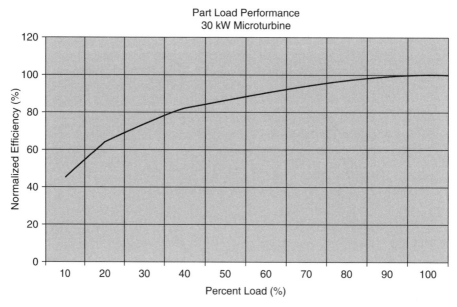

FIGURE 3-3 *Microturbine part-load efficiency. Note: Unit represented is a single-shaft, high-speed alternator system. (Source: Gas Research Institute and the National Renewable Energy Laboratory. 2003.* Gas-fired distributed energy resource technology characterizations. *U.S. Department of Energy. Oak Ridge, TN. Taken from: Energy and Environmental Analysis, Inc., estimates.)*

sea level and 59°F. Note that some manufacturers may place limitations on maximum power output below a certain ambient temperature due to maximum power limitations of the gearbox, generator, or power electronics, and this may modify the shape of the curves shown in Figure 3-4.

The density of air changes with altitude. Density decreases at increasing altitudes, and, consequently, power output decreases. Figure 3-5 illustrates the typical derating effect of altitude on microturbine power output.

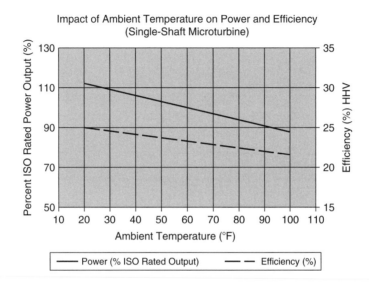

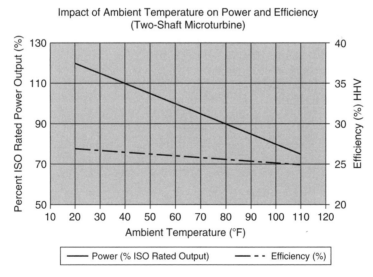

FIGURE 3-4 *Ambient temperature effects on microturbine performance. Note: Power presented as a percent of International Organization for Standards (ISO)-rated power over range of ambient conditions; efficiency represented as net electric efficiency over ambient conditions. (Source: Gas Research Institute and the National Renewable Energy Laboratory. 2003.* Gas-fired distributed energy resource technology characterizations. *U.S. Department of Energy. Oak Ridge, TN.)*

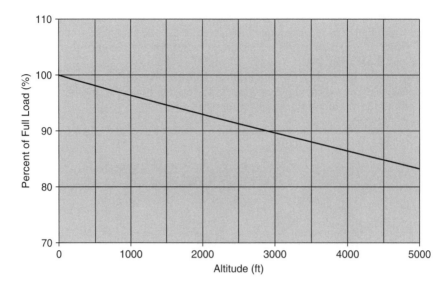

FIGURE 3-5   *Altitude effects on microturbine power output. (Source: Gas Research Institute and the National Renewable Energy Laboratory. 2003. Gas-fired distributed energy resource technology characterizations. U.S. Department of Energy. Oak Ridge, TN. Taken from: Energy and Environmental Analysis, Inc., estimates.)*

COMBINED HEAT AND POWER PERFORMANCE

Use of the thermal energy contained in the exhaust gas improves microturbine system economics. The heat in the exhaust represents close to 70% of the inlet fuel energy. Exhaust heat can be recovered and used in a variety of ways, the most common being to generate hot water for water heating, space heating, or process uses or for "driving" thermally activated equipment such as absorption chillers or desiccant dehumidifiers.

Microturbine CHP system efficiency is a function of the exhaust gas temperatures from the recuperator and from the heat-recovery unit. Manufacturers use recuperators of varying effectiveness in an effort to balance microturbine electricity-generating efficiency, CHP efficiency, and equipment cost. Since manufacturers have developed substantially different designs for recuperators, their effectiveness varies among manufacturers and systems.

Recuperator effectiveness strongly influences the microturbine exhaust temperature. Consequently, the various microturbine CHP systems have different CHP efficiencies and net heat rate chargeable to power. Variations in efficiency and net heat rate chargeable to power are mostly due to the mechanical design and manufacturing cost of the recuperators and their resulting impact on system cost, rather than being a function of system size.

Chapter 4

Microturbine Economics and Market Factors

This chapter discusses estimates for the installed cost of microturbine systems. Two configurations are studied:

1. Power only
2. Combined heat and power (CHP), producing hot water by using waste heat recovery

Original equipment manufacturer (OEM) package costs and installed costs are discussed. Note that costs vary considerably depending on the following:

- Choice of OEM and OEM design
- Geographical area and other demographics
- Market conditions
- Special site requirements
- Emissions control requirements
- Prevailing labor rates
- Whether the distributed generation system is a new or retrofit application

One item from the author's notes at the front of the book bears repeating as follows: Any cost values of this or similar technologies will vary according to technological progress, manufacturers' infrastructure, fuel prices both global and national, and therefore the country the work is done in. So all cost values need to be checked again by the reader, as required. Cost data presented however, does serve to provide a relative cost comparison between different technologies.

SAMPLE COST DATA: MICROTURBINE CHP SYSTEMS[*]

Table 4-1 provides cost estimates for CHP applications. The basic microturbine package consists of the turbogenerator package and power electronics. All of the commercial and near-commercial units offer basic interconnection and parallel functions as part of the package cost, but utilities typically require additional safety and control hardware for interconnection.

All but one of the systems offers an integrated heat exchanger/heat-recovery system for CHP within the package. Three of the most prominent current manufacturers offer an integral gas booster compressor as an option to the base package price. One manufacturer includes the gas booster compressor within the base package price. Note that the package and component prices cited in Table 4-1 represent manufacturer quotes

---

[*] Adapted from: Gas Research Institute and the National Renewable Energy Laboratory. 2003. *Gas-fired distributed energy resource technology characterizations*. U.S. Department of Energy. Oak Ridge, TN.

TABLE 4-1   Estimated Capital Cost for Typical Microturbine Generator Systems in Grid-Interconnected Combined Heat and Power Applications (2003)

Cost Component	System 1	System 2	System 3	System 4
Nominal capacity (kW)	30	70	80	100
Cost ($/kW)				
Equipment				
Microturbine package	1100	1070	1000	920
Gas booster compressor	180	included	75	75
Heat recovery	100	included	included	included
Add'l controls/interconnect	180	115	115	100
Total equipment	1560	1185	1190	1095
Labor/materials	655	444	445	403
Total process capital	2215	1629	1635	1498
Project and construction and management	156	119	119	108
Engineering and fees	156	95	95	88
Project contingency	119	83	83	75
Total plant cost (2003 $/kW)	$2636	$1926	$1932	$1769

(Source: Gas Research Institute and the National Renewable Energy Laboratory. 2003. *Gas-fired distributed energy resource technology characterizations*. U.S. Department of Energy. Oak Ridge, TN. Taken from: Energy and Environments Analysis, Inc., estimates based on manufacturer equipment cost estimates.)

or estimates current at the time these quotes were made. Note that prices may vary depending on breakthroughs in technology, changes in associated factors, and costs of supporting technology and construction. For instance, if tariffs for conventional power were to rise, by say as much as 100%, based on rocketing fuel prices, manufacturers of distributed power packages, including microtubine OEMs, would then feel justified in raising their prices, even if their own costs did not change much.

Manufacturer quotes may not reflect actual manufacturing cost plus profit, but may instead represent a forward *pricing* strategy in which early units are sold at a loss to develop the market.

SAMPLE COST DATA: MICROTURBINE POWER-ONLY SYSTEMS

The information provided for four power-only sample systems (accurate at the time the information was compiled) in power-only (non-CHP service) is as follows:

System 1. 30 kilowatts (kW): Single unit $1100/kW, direct current (DC)–to–alternating current (AC) inverter, basic electronic interconnection hardware, but without a heat recovery heat exchanger. Additional cost is involved for integral fuel gas compressor. Prices are expected to be lower for volume purchases.

System 2. 70 kW: A price of $1070/kW includes integral heat recovery heat exchanger, fuel gas compressor, and generator with basic controls and interconnection hardware (standard 3600-RPM AC generator).

System 3. 80 kW: A price of $1000/kW includes integral heat recovery heat exchanger, DC-to-AC inverter, and basic electronic interconnection hardware. Additional costs are involved for integral fuel gas compressor.

<u>System 4.</u>   100 kW: A price of $920/kW includes integral heat recovery heat exchanger, DC-to-AC inverter, and basic interconnection hardware. Additional costs are involved for integral fuel gas compressor.

The equipment cost represents the price for the ultimate customer that will include the microturbine package and other major components. The cost for a customer installing a microturbine-based CHP system includes a number of other factors that increase the total costs by an estimated 60% to 70% over equipment costs.

The total installed-cost estimates are based on a simple installation with minimal site preparation required. These cost estimates include provisions for grid interconnection and paralleling. Labor and materials represent the labor cost for the civil, mechanical, and electrical work and materials such as ductwork, piping, and wiring and is estimated to range from 35% to 45% of the total equipment cost.

The equipment costs—plus installation labor and materials—is defined as *total process capital*. A number of other costs are incurred on top of total process capital. These costs are often referred to as *soft costs* because they vary widely by installation, development channel, and approach to project management.

Project and construction management also includes general contractor markup and bonding (and performance guarantees) and is estimated to be 10% of the total equipment cost. Engineering and permitting fees are required to design the system and integrate it functionally with the application's electrical and mechanical systems and are estimated to range from 8% to 10% of the total equipment cost, depending on the system size. Contingency is assumed to be 5% of the total equipment cost in all four systems.

The capital costs will be lower for power-only systems that have no heat-recovery equipment. For the units that integrate this equipment into the basic microturbine package (i.e., the 70-, 80-, and 100-kW systems), savings are projected at approximately

TABLE 4-2 Estimated Capital Cost for Typical Microturbine Generator Systems in Grid-Interconnected Power-Only Applications (2003)

Cost Component	System 1	System 2	System 3	System 4
Nominal capacity (kW)	30	70	80	100
Cost ($/kW)				
Equipment				
Microturbine package	1100	1070	1000	920
Gas booster compressor	180	included	75	75
Heat recovery	n/a	n/a	n/a	n/a
Controls/monitoring	180	115	115	100
Total equipment	$1460	$1185	$1190	$1095
Labor/materials	438	261	261	240
Total process capital	$1898	$1446	$1451	$1335
Project and construction management	146	114	114	105
Engineering and fees	146	91	91	84
Project contingency	73	57	57	52
Total plant cost (2003 $/kW)	$2263	$1708	$1713	$1576

(Source: Gas Research Institute and the National Renewable Energy Laboratory. 2003. *Gas-fired distributed energy resource technology characterizations*. U.S. Department of Energy. Oak Ridge, TN. Taken from: Energy and Environments Analysis, Inc., estimates based on manufacturer equipment cost estimates.)

$50/kW if heat recovery equipment is not installed. In addition, installation labor and materials costs are reduced (estimated to be 25% to 30% of total equipments costs) because there is no need to connect heat-recovery equipment to the application's thermal system.

Engineering and permitting fees required to design the system and integrate it functionally with the application's electrical and mechanical systems are estimated to range from 8% to 10% of the total equipment cost, depending on the system size. Project management and construction fees remain 10% of total equipment costs. Contingency for power-only installations is 5% due to reduced complexity and risk in the installation. Table 4-2 shows the cost estimates for power-only versions of the four microturbine systems.

As an emerging product, the capital costs shown in Tables 4-1 and 4-2 represent the cost to the customer for early-market-entry product, but not the cost of the very first units into the market. All of the microturbine developer/manufacturers have cost-reduction plans and performance-enhancing developments for future, more mature products.

CASE STUDY 4-1: DETERMINING PROJECT ECONOMICS AND ROI

Demographic factors, geographical locations, and market and political conditions may drastically affect factors such as fuel prices. This can change the relative importance of the parameters and variables discussed in the following example.

Figures 4-1 and 4-2 (details used to construct Figure 4-1) illustrate the first step in determining the usefulness of a distributed generation installation. Figure 4-1 demonstrates an end user's peak demand of 92 kW reduced to 46 kW (peak load shaving) with the installation of a microturbine. Using data applicable to this end user, power supply company (Boston Edison) data, and electricity prices current at that time, it can be noted

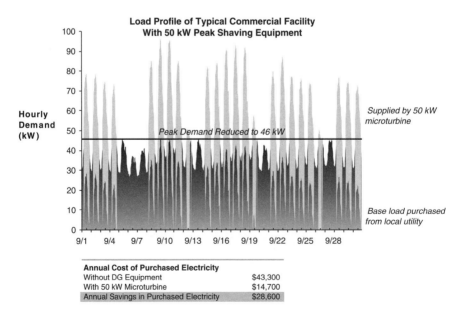

FIGURE 4-1 *Distributed generation (DG) project example. (Source: Arthur D. Little. 1999. Distributed energy: understanding the economics. Arthur D. Little, Inc. Cambridge, MA.)*

Grid Cost of Delivered Energy		
Boston Edison's G-2 Rate (10–200 kW)	**Oct.–May**	**June–Sept.**
Demand Charges ($/kW) (in excess of 10 kW)	10.54	22.59
Energy Charges (¢/kWh) (includes DSM and renewables)	4.10	4.10
Distribution Charges		
First 2000 kWh (¢/kWh)	1.1	2.1
Next 150 kWh (¢/kWh)	.6	.8
Additional kWh (¢/kWh)	.5	.5
Transition Charges		
First 2000 kWh (¢/kWh)	4.3	8.4
Next 150 kWh (¢/kWh)	2.2	2.9
Additional kWh (¢/kWh)	1.4	1.6

	Annual Cost of Electricity Purchased From Utility				
	Peak Demand[1] (kw)	Demand Charges ($)	Energy[2] (kWh)	Energy Charges ($)	Total Cost ($)
Without DG Equipment	75–95	13,700	439,000	29,600	43,300
With 50 kW Microturbine	25–46	5000	125,000	9700	14,700
Annual Savings in Purchased Electricity		8700		19,900	28,600

1. Varies by month
2. Includes energy, DSM, renewables, distribution, and transition charges

FIGURE 4-2   *Grid cost of delivered energy. [1]Varies by month. [2] Includes energy, DSM, renewables, distribution, and transition charges. DG = distributed generation; DSM = demand side management. (Source: Arthur D. Little. 1999.* Distributed energy: understanding the economics. *Arthur D. Little, Inc. Cambridge, MA.)*

that the customer's purchased power costs drop from $43,300 annually to $14,700, a numerical savings of $28,600.

To estimate actual savings, added monetary and nonmonetary benefits (such as convenience, some independence from a power company, flexibility of operation, and electric distribution company benefits) need to be considered. Additional costs (such as capital costs incurred by the microturbine installation) must also be factored in to help complete the economic picture. Distributed generation operating costs (fuel costs plus overhaul and maintenance costs) must be subtracted from the power costs savings (see previous paragraph) to arrive at the figure for total annual savings (Figure 4-3).

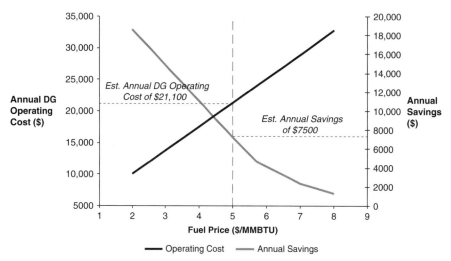

FIGURE 4-3 *Distributed generation (DG) operating cost and annual savings versus fuel price. O&M = overhaul and maintenance; LHV = lower heating value. (Source: Arthur D. Little. 1999.* Distributed energy: understanding the economics. *Arthur D. Little, Inc. Cambridge, MA.)*

Chapter 5

Microturbine Fuels and Emissions

FUELS*

Microturbines have been designed to use natural gas as their primary fuel. However, most models can operate on a wide variety of fuels, including the following:

- Liquid fuels (e.g., distillate oil)
- Liquefied petroleum gas: Propane and butane mixtures
- Sour gas: Unprocessed natural gas; comes directly from some gas wells
- Biogas: Any of the combustible gases produced from biological degradation of organic wastes, such as landfill gas, sewage digester gas, and animal waste digester gas
- Industrial waste gases: Flare gases and process off-gases from refineries, chemical plants, and steel mills
- Manufactured gases: Typically low- and medium-BTU gas produced as products of gasification or pyrolysis processes
- Biomass, treated

Contaminants are problematic with most waste fuels, specifically acid gas components (H_2S, halogen acids, HCN [from ammonia, salts, and metal-containing compounds], organic halogen-, sulfur-, nitrogen-, and silicon-containing compounds); and oils. In combustion, halogen and sulfur compounds form halogen acids, SO_2, sonic SO_3, and possibly H_2SO_4 emissions.

The acids can also corrode downstream equipment. A substantial fraction of any fuel-bound nitrogen will be oxidized into NO_x in combustion. Particulates (and liquids, in the case of gaseous fuels) must be kept to very low concentrations to prevent corrosion and erosion of components. Various fuel-scrubbing, droplet-separation, and filtration steps will be required if any fuel contaminant levels exceed manufacturer specifications.

Landfill gas, in particular, often contains chlorine compounds, sulfur compounds, organic acids, and silicon compounds, which necessitate pretreatment.

EMISSION CHARACTERISTICS

Microturbines have the potential for extremely low emissions. Most current micro-turbines operating on gaseous fuels feature lean premixed (dry low NO_x [DLN])

---

* Adapted from: Gas Research Institute and the National Renewable Energy Laboratory. 2003. *Gas-fired distributed energy resource technology characterizations*. U.S. Department of Energy. Oak Ridge, TN.

combustor technology, which was developed relatively recently for gas turbines but is not universally featured on larger gas turbines. Because microturbines are able to meet key emissions requirements with this or similar built-in technology, post-combustion emission control (after-treatment) techniques are currently not needed.

The primary pollutants from microturbines are oxides of nitrogen (NO_x), carbon monoxide (CO), and unburned hydrocarbons (THC). Microturbines emit a negligible amount of sulfur dioxide (SO_2) depending on the amount of sulfur in the fuel. Microturbines using lean premix combustion are designed to achieve the objective of low emissions at full load. Emissions are often higher when operating at part load.

Emission Control Options

Lean Premixed Combustion

Thermal NO_x formation is a function of both the local temperatures within the flame and residence time. In older combustors used in industrial gas turbines, fuel and air were separately injected into the flame zone. Such separate injections resulted in high local temperatures where the fuel and air zones intersected, and most of the burning took place at stoichiometric conditions. *Stoichiometric conditions* are when the amount of oxygen in the air is sufficient for combustion of the fuel and there is no unused fuel or oxygen after combustion.

The focus of combustion improvements of the past decade was to lower the flame temperature of local hot spots using lean fuel/air mixtures, so zones of high local combustion temperatures were not created. Lean combustion decreases the fuel-air ratio in the zones where NO_x production occurs, so that peak flame temperature is less than the stoichiometric flame temperature, thereby suppressing thermal NO_x formation.

Most microturbines now feature lean premixed combustion systems, also referred to as dry low NO_x (DLN) or dry low emissions. Lean, premixed combustion mixes the gaseous fuel and compressed air before they enter the combustion zone so that there are no local zones of high temperatures, or "hot spots," where high levels of NO_x would form.

DLN requires specially designed mixing chambers and mixture inlet zones to avoid flashback of the flame. Optimized application of DLN combustion requires integrated combustor and turbine design techniques, and specific combustor designs must be developed for each turbine application.

NO_x emissions below 9 parts per million by volume (ppmv) at 15% O_2 have been achieved with lean premixed combustion in microturbines. Most manufacturers guarantee NO_x emissions of 9 to 25 ppmv at full-load operation in their product introductions. Part-load emission levels can be significantly higher and are an area of continued development activities.

Catalytic Combustion

In catalytic combustion, fuel is oxidized at low temperatures (typically below 1700°F, where NO_x formation is very low) in the presence of a catalyst. The catalyst is applied to a high-surface-area substrate and causes a fuel-air mixture to react and release its initial thermal energy.

This initial energy (heat) stabilizes gas-phase combustion at moderate flame temperatures (as low as 2200°F). Additional air or fuel can be introduced downstream of the catalyst to complete combustion and/or to achieve the desired combustor outlet design temperature. By enabling such stable low-temperature combustion, catalytic combustors emit less than 3 ppm NO_x—with low CO and UHC emissions and with low levels of vibration and acoustic noise.

Combustion system and gas turbine developers, along with the U.S. Department of Energy, the California Energy Commission, and other government agencies, are continuing to develop gas turbine catalytic combustion technology. Past efforts at developing catalytic combustors for gas turbines, achieved low, single-digit NO_x ppm levels but failed to produce combustion systems with suitable operating durability.

This was mostly due to cycling damage and to the brittle nature of the materials used for catalysts and catalyst support systems. Catalytic combustor developers and gas turbine manufacturers are testing durable catalytic and "partial catalytic" systems that are overcoming the problems of past designs.

Catalytic combustors capable of achieving NO_x levels below 3 ppm are in full-scale demonstration and are entering early commercial introduction. As with DLN combustion, optimized catalytic combustion requires an integrated approach to combustor and turbine design.

Catalytic combustors must be tailored to the specific operating characteristics and physical layout of each turbine design. Catalytic combustion may be applied to microturbines as well as industrial and utility turbines. For example, Kawasaki offers a version of their MIA 13X, 1.4-MW gas turbine with a catalytic combustor, with less than 3 ppm NO_x guaranteed.

SYSTEM EMISSIONS

Table 5-1 shows typical emissions for microturbine systems. The data shown reflect manufacturers' guaranteed levels.

Environmental Degradation Caused by Water Vapor Emissions

A prevalent recuperator operational problem is environmental degradation caused by water vapor. The following paragraphs describe the degradation mechanisms and experimental observations made in one original equipment manufacturer's study (see Chapter 6, Performance Optimization and Testing, for further details).

TABLE 5-1   Microturbine Emission Characteristics

Emissions Characteristics	System 1	System 2	System 3	System 4
Nominal capacity (kW)	30	70	80	100
Electrical efficiency (%), HHV	23	25	24	26
NO_x (ppmv at 15% O_2)	9	9	25	15
NO_x, lb/MWh[*]	0.51	0.45	1.25	0.72
CO (ppmv)	40	9	50	15
CO (lb/MWh)	1.38	0.27	1.51	0.45
THC (ppmv)	<9	<9	<9	<9
THC (lb/MWh)	<0.18	<0.16	<0.16	<0.15
CO_2 (lb/MWh)	1765	1585	1650	1535

HHV = higher heating value; MWh = megawatt hours.
Note: Estimates are based on manufacturers' guarantees for typical systems commercially available in 2003; emissions estimates represent 15% O_2 using natural gas fuel.
[*]Conversion from volumetric emission rate (ppmv at 15% O_2) to output-based rate (lbs/MWh), for both NO_x and CO, was based on conversion multipliers provided by Catalytica Energy Systems.
(Source: Gas Research Institute and the National Renewable Energy Laboratory. 2003. *Gas-fired distributed energy resource technology characterizations*. U.S. Department of Energy. Oak Ridge, TN.)

A prevalent recuperator operational problem is environmental degradation caused by water vapor. The following two paragraphs describe the degradation mechanisms and general observations made in experimental studies.

Water[*] vapor is encountered as a minor component in ambient air and in larger concentrations as a by-product of combustion processes. It has been known for some time that the presence of water vapor in oxidizing environments can alter the degradation process for many different metals.

When present in oxygen-bearing atmospheres or as the primary oxidant, water vapor appears to hasten the onset of rapid oxidation of Fe-Cr and Fe-Ni-Cr alloys at elevated temperatures. The results vary from study to study, but general trends show that the presence of water vapor accelerates the rate of oxidation, leads to the formation of layered scales and increases the amount of chromium required to form a protective oxide film. Work on austenitic stainless steels suggests that high chromium and nickel levels are beneficial.

Recent studies involving Type 347 austenitic stainless steel foil noted that a thin chromium oxide layer is established on the sample surface during initial exposure to air containing water vapor. The amount of chromium included into this scale is not significant enough to result in breakaway oxidation due solely to chromium depletion of the substrate.

Rapid weight gain then occurs due to the formation and growth of mixed oxide nodules. These nucleate after an incubation time and then spread, consuming the initially formed protective oxide layer. This is corroborated by the fact that remnants of the initial chromium oxide layer survive on heavily degraded samples. It is unclear if the nodules form as a direct result of the action of water vapor or if it plays a role in inhibiting healing of flaws, cracks, and spalled regions by the formation of new chromium oxide. One mechanism proposed for the actions of water vapor on several systems is the increase in the rate of formation of volatile compounds.

Cr_2O_3 evaporates in the absence of water as higher chromium oxides such as CrO_3, an effect most pronounced temperatures above about 1000°C (1832°F). Water vapor can increase the evaporative heat loss to levels where it is at significantly lower temperatures, particularly in rapidly flowing gas streams, due to the formation of volatile oxy-hydroxides such as $CrO_2(OH)_2$.

[*] Source: Montague, J. P., Stinner, C., Rakowski, J. M., and Lipschutz, M. D. 2004. The use and performance of oxidation and creep resistant stainless steels in an exhaust gas primary surface recuperator application. *Proceedings of ASME Turbo Expo 2004.* American Society of Mechanical Engineers. Vienna, Austria.

Chapter 6

Microturbine Performance Optimization and Testing

This chapter describes state-of-the-art technology improvements that manufacturers, researchers, and government officials are pursuing that are expected to result in enhanced microturbine performance for the mid term (the next 15 to 20 years). Most of them will also pave the way for long-term improvements.

TECHNOLOGY PATHS TO INCREASED PERFORMANCE

Turbine inlet temperature is the main parameter that large gas turbine designers focus on in gas turbine technology advancement efforts. Increases in turbine inlet temperature rapidly increase the power output of the turbine—and, to a lesser extent, increase efficiency. With an increase in firing temperature, a corresponding increase in pressure ratio yields heightened benefits.

Modern large gas turbines use sophisticated methods of internal turbine cooling to permit higher temperatures without exceeding the turbine material's metallurgical creep limits. Microturbines, however, presently cannot take advantage of internal turbine cooling because the complex shape of the flow passages of small radial turbines does not yet lend itself to cost-effective manufactured configurations.

Internal cooling in microturbines also faces the problem of an expansion process involving a single turbine stage. Single-stage turbines result in cooling air being unavailable for subsequent power generation; cooling air cannot enter later turbine stages and provide additional power as is the case with multistage, axial turbines. Such diversion of pressurized air for cooling purposes results in efficiency and power penalties. Material advances have therefore become the preferred route to the higher temperatures that would increase microturbine efficiencies.

Calculated estimates for potential efficiency and specific power of microturbines with ceramic turbines, combustor, and associated high-temperature components are shown in Figures 6-1 and 6-2. These figures include the performance of microturbines with metallic turbines and firing temperatures of 1550°F to 1750°F for comparison. The performance of microturbines with ceramic components is shown over the temperature range of 2100°F to 2500°F. The lowest temperature for which engineering development is worthwhile pursuing is in the 1900°F to 2100°F range.

Any move to use firing temperatures above 2500°F is not likely until microturbines with firing temperatures below this level have proved themselves in service. At a 2500°F firing temperature, the expected efficiency of a microturbine would be just above the 40% (lower heating value) level.

Indications are that ceramics have the potential for a significant performance gain in terms of both efficiency and specific power which, if successfully developed and deployed, would enhance the economics of microturbine systems. At some point around

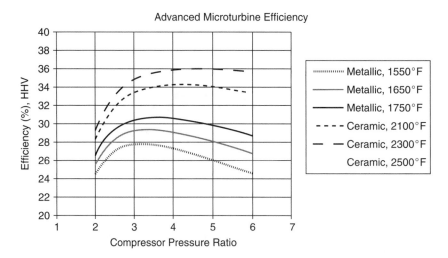

FIGURE 6-1   *Efficiency comparison of microturbines with metallic and ceramic turbines. (Source: Gas Research Institute and the National Renewable Energy Laboratory. 2003.* Gas-fired distributed energy resource technology characterizations. *U.S. Department of Energy. Oak Ridge, TN.)*

the range of 2400°F to 2500°F, microturbine emissions may reach a limit whereby further advances are not practical without some other combustion process (e.g., catalytic combustion) or other novel emissions suppression technique.

Increasing turbine inlet temperature by using either ceramic materials or internal turbine cooling would also increase turbine exhaust temperature and the recuperator

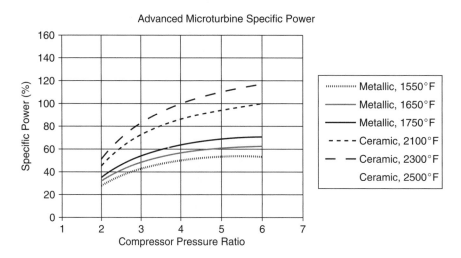

FIGURE 6-2   *Specific power comparison of microturbines with metallic and ceramic turbines. (Source: Gas Research Institute and the National Renewable Energy Laboratory. 2003.* Gas-fired distributed energy resource technology characterizations. *U.S. Department of Energy. Oak Ridge, TN.)*

inlet temperature. In current microturbines, the recuperator inlet temperature is at, or very near, the metallurgical limit of type AISI 347 stainless steel, the generally preferred recuperator material.

Although alloys with higher temperature capabilities are available commercially (such as Inconel 625), such alloys are much more expensive. A forced switch to such alloys could increase microturbine cost enough to negate the value of any efficiency gains. Several alternatives are available to circumvent this potential recuperator material problem. Turbine exhaust/recuperator inlet temperature can be reduced by increasing turbine pressure ratio.

Pressure ratios that are only slightly higher than the optimum for maximum efficiency compromise efficiency only to a small extent. Additionally, slightly higher pressure ratios than those required for maximum efficiency increase specific power so that the economics of the equipment by itself may improve due to the increased power output and reduction of capital costs per net kilowatt (kW).

The performance of internally cooled metallic turbines and ceramic turbines are similar enough that performance predictions for future ceramic turbines can be represented by those of internally cooled metallic turbines. Performance predictions for internally cooled radial inflow turbines for microturbine application would be generally similar to those for ceramic turbines but would require slightly higher turbine inlet temperature to compensate for the performance sacrifice due to the use of a small amount of compressed air for cooling. However, internally cooled radial inflow turbines are not yet available, so an accurate estimate of cooling air requirements is not available at this time.

Only a limited reduction of turbine exit temperature can be accomplished by increasing the turbine pressure ratio. This limitation is due to aerodynamics, stresses resulting from centrifugal forces, and materials limitations.

When greater reductions in turbine exit/recuperator inlet temperature are required, then microturbine system designers may elect to go to entirely new cycle designs such as an intercooled recuperated (ICR) cycle, with its substantially higher pressure ratio and consequentially lower recuperator inlet temperature. A microturbine ICR cycle would employ two compressors, most likely both centrifugal, which would result in a pressure ratio of 10 to 15.

Small aeronautical propulsion gas turbines have been built with two-stage centrifugal compressors. With a two-stage compressor, it would be straightforward to add an intercooler, possibly of a design and construction of the type used in turbocharged truck or marine power applications.

The ICR cycle offers greater specific power as well as the ability to go to higher turbine inlet temperature without raising the recuperator inlet temperature above the capability of type AISI 347 stainless steel. Such a cycle change would present a discontinuity in efficiency and power increase and would require some machine development. It should be considered an alternative path to increased performance and reduced fired cost per kW.

CERAMIC MATERIALS

Because microturbines are just now entering the market, there has been inadequate time and market pull for an advanced technology component business to develop. Thus, the U.S. Department of Energy (DOE) has begun an advanced materials program for microturbines to develop ceramic turbines, combustors, recuperators, and associated high-temperature components.

Ceramics allow significantly higher firing temperatures, which will enable microturbines to achieve much higher efficiencies and specific power ratings. It is unlikely that advancing the temperature of microturbines will be a slow, incremental process. Ceramic materials temperature capabilities are much higher than those of (uncooled) metallic turbines. When ceramic materials have been fully developed and proven practical for commercial turbine service, microturbine performance—in terms of both efficiency and specific power—will take a substantial step forward.

Monolithic structural ceramics for gas turbine applications have matured greatly during the past 35 years. The ceramic materials available at the beginning of the 1970s in the United States were reaction-bonded silicon nitride, hot-pressed silicon nitride, and reaction-sintered silicon carbide, primarily based on technology developed in Great Britain during the 1960s.

These materials were relatively weak (particularly at elevated temperatures), had low fracture toughness, and were subject to time-dependent failure (i.e., creep rupture and slow crack growth). Continued improvements in powder synthesis, processing, and densification have resulted in the development of a current generation of silicon nitride ceramics having controlled microstructures consisting of elongated grains. These so-called self-reinforced or in-situ toughened materials exhibit superior performance as reflected by increased strength, higher fracture toughness, and enhanced resistance to creep rupture (lifetimes increased by four orders of magnitude).

Similar progress has been made in the areas of (1) component design and fabrication, (2) probabilistic design methodologies for predicting component life, and (3) nondestructive evaluation of complex-shaped parts. Because of these developments, monolithic ceramic components (blades and vanes) have accumulated thousands of hours under tests in recent gas turbine engine demonstration programs. Despite the experience base described above, a number of challenges remain before monolithic and composite ceramics exhibit the long lifetimes required for turbine applications. The first challenge is that of environmental degradation.

The above-mentioned field tests have shown that three processes can affect performance of the monolithic ceramics: (1) localized corrosion due to the presence of reactive species in the environment, (2) environmentally induced destabilization of the intergranular phases, and (3) rapid recession of the silicon nitride (or silicon carbide) due to loss of the silica scale by direct reaction with water vapor. This has led to the development of oxide-based environmental barrier coatings, which are at an early stage of development and are the subject of substantial rig and engine testing.

The second challenge, which is specific for monolithic ceramics, is that of impact resistance. Further improvements in fracture toughness may be required to ensure high reliability in the event of foreign object damage.

As noted earlier, increasing firing temperatures also increases the temperature of the hot exhaust to the recuperator. Firing temperatures above 2000°F to 2100°F probably would require a ceramic recuperator. The development of low-cost, reliable ceramic recuperators would require a separate, intensive research and development effort in parallel to the ceramic turbine program.

Case 6-1[*] describes some results obtained with 100-micron wafers of nickel chrome alloy with niobium in simulated recuperator service at temperatures of between 700°C and 760°C.

Case 6-2[**] details experimental techniques, oxidation mechanisms, and results experienced by recuperator test wafers in service temperatures as high as 800°C with

[*] Adapted extracts from: Rakowsky, J. M., Stinner, C. P., Lipshutz, M., and Montague, J. P. 2004. The use and performance of oxidation and creep-resistant stainless steels in an exhaust gas primary surface recuperator application. *Proceedings of ASME Turbo Expo 2004*. American Society of Mechanical Engineers. Vienna, Austria.

[**] Adapted extracts from: Pint B. A. and More K. L. 2004. Stainless steels with improved oxidation resistance for recuperators. *Proceedings of ASME Turbo Expo 2004*. American Society of Mechanical Engineers. Vienna, Austria.

various recuperator alloys. (The complete study also collected data and lower service temperature ceilings.)

Case 6-3[***] provides more details than the previous case on recuperator alloy types and compositions, as well as their market readiness and current costs.

Case 6-4[^] describes some of the work in a Chinese university study on a commercially available microturbine. The interesting aspects of this case include a discussion of a rotating disk regenerator (possibly better heat transfer potential than a stationary regenerator if substantial air flow leaks can be avoided) and a manually controlled start sequence that explores the various speed thresholds in a start cycle. The Chinese were also testing commercially available software to explore its potential use in further tests.

CASE 6-1[*]

Case 6-1 describes some results obtained with 100-micron wafers of nickel chrome alloy with niobium in simulated recuperator service at temperatures of between 700°C and 760°C.

Test samples of 20-25+Nb stainless steel were prepared by shearing from a wrought coil of 100-micron-thick (0.004-inch) foil. The source of the oxidation test samples was the pilot-scale coil melted at ATI Allvac. The mill surface, which was produced by cold rolling followed by bright annealing in dry hydrogen, was degreased but was not otherwise altered prior to the oxidation test exposures.

The oxidation test protocol used for this study was cyclic in nature, with an average period of approximately 150 hours between thermal cycles. The samples were rapidly air-cooled to room temperature within 15 minutes. Several identical samples were exposed at each test condition. Some of these samples were removed at different total exposure times for metallographic evaluation, whereas others were included for the entire duration of the test.

Ambient air exposures were used as a baseline. Samples were exposed in alumina crucibles in a box furnace. Gas flow through the furnace chamber was due solely to convection. The typical humidity in the ambient laboratory environment corresponds to approximately 0.5% (by volume) water vapor. Emphasis was placed on exposures at 704°C (1300°F) and 760°C (1400°F), with test periods ranging from 500 to 10,000 hours. The results are shown in Figure 6-3. Note that the two solid lines are idealized weight change curves calculated from the parabolic rate constants (kp) obtained from the individual tests. The weight-change data indicate that the terminal oxide thickness should be about 2 to 3 microns.

In summary, a 20Cr-25Ni austenitic stainless steel stabilized with niobium was tested for suitability for high-temperature service, with particular emphasis placed

[***] Adapted extracts from: Maziasz P. J., Pint, B. A., Shingledecker, J. P., More, K. L., Evans, N. D., and Lara-Curzio, E. 2004. Austenitic stainless steels and alloys with improved high-temperature performance for advanced microturbine recuperators. *Proceedings of ASME Turbo Expo 2004*. American Society of Mechanical Engineers. Vienna, Austria.

[^] Adapted extracts from: Chiang, H.-W. D., Wang, C.-H., and Hsu, C.-N. 2004. Performance testing of a microturbine generator set with twin rotating disk regenerators. *Proceedings of ASME Turbo Expo 2004*. American Society of Mechanical Engineers. Vienna, Austria.

[*] Adapted extracts from: Rakowsky, J. M., Stinner, C. P., Lipshutz, M., and Montague, J. P. 2004. The use and performance of oxidation and creep-resistant stainless steels in an exhaust gas primary surface recuperator application. *Proceedings of ASME Turbo Expo 2004*. American Society of Mechanical Engineers. Vienna, Austria.

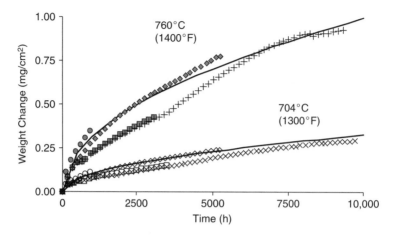

FIGURE 6-3   *Specific weight change data for oxidation of a 20-25+Nb alloy in ambient air at 704°C and 760°C. Solid lines indicate idealized fit to average experimentally determined rate constant values. (Source: Rakowsky, J. M., Stinner, C. P., Lipshutz, M., and Montague, J. P. 2004. The use and performance of oxidation and creep-resistant stainless steels in an exhaust gas primary surface recuperator application. Proceedings of ASME Turbo Expo 2004. American Society of Mechanical Engineers. Vienna, Austria.)*

on usage in a primary surface recuperator application. This alloy, in the form of 100-micron-thick foil, was found to have excellent oxidation resistance in humidified air for times as long as 10,000 hours at temperatures as high as 760°C (1400°F). Oxidation failure was not observed during the test program, indicating that the oxidation lifetime limit is likely to be at even longer times and/or higher temperatures.

The alloy appears to be relatively stable in terms of microstructure, with no sigma or other brittle phases noted for exposure times as long as 10,000 hours. The measured resistance to creep deformation places this alloy approximately intermediate in creep strength between the common austenitic stainless steels such as types 347 and 310 and the high-strength solid solution strengthened nickel-base alloys, e.g., 625 and HX alloys (UNS N06625 and N06002, respectively). Finally, this alloy was found to be fully compatible with standard methods and practices of recuperator fabrication.

CASE 6-2[**]

Case 6-2 details experimental techniques, oxidation mechanisms, and results experienced by recuperator test wafers in service temperatures as high as 800°C with various recuperator alloys. (The complete study also collected data and lower service temperature ceilings.)

The performance of two groups of alloys will be presented in this case study. The first group includes more highly alloyed commercial alloys, an Fe-base alloy, 120; and a Ni-base alloy, 625. These alloys are three to five times more expensive than type 347 stainless steel. The second group comprises stainless steels, not currently produced commercially as foil, that have slightly higher Cr and Ni contents than type

[**] Adapted extracts from: Pint B. A. and More, K. L. 2004. Stainless steels with improved oxidation resistance for recuperators. *Proceedings of ASME Turbo Expo 2004.* American Society of Mechanical Engineers. Vienna, Austria.

347 stainless steel. These alloys are hoped to result in a less than 50% higher material cost. As expected, the first group has better corrosion resistance than the second group.

However, because cost is a major issue with the implementation of microturbines, the second group may be key to allowing recuperator operating temperatures to move up to 700°C. Thus, this group is the main focus of the Oak Ridge National Laboratory (ORNL) alloy development program, which is now moving into mill-scale foil production.

Experimental Procedure

The chemical compositions of selected alloys examined in this study are listed in Table 6-1. Some of the materials were obtained from commercial vendors in foil form whereas others were obtained in thicker sections and then hot- and cold-rolled at ORNL to approximately 100-micron thickness with average grain sizes given in Table 6-1.

Model alloys were vacuum-induction melted at ORNL and hot- and cold-rolled to a 1.25-mm sheet. After the final cold-rolling step, the sheets were annealed in Ar for 2 minutes at 1000°C. Selected alloys were then rolled to foil under similar conditions as those used for the commercial alloys.

Foil specimens (approximately $12 \times 18 \times 100$ microns) were tested in the as-rolled conditions, and similar-sized sheet specimens were polished to a 600 SiC grit finish. The specimens were cleaned in acetone and methanol prior to oxidation, and mass changes were measured using a Mettler-Toledo model AG245 balance. Exposures were 100-hour cycles at 650°, 700°, or 800°C.

Oxidation exposures in humid air were conducted by flowing the gas at 450 cc/min through an alumina tube that was inside a resistively heated tube furnace. Distilled water was atomized into the flowing gas stream above its condensation temperature and heated to the reaction temperature within the alumina tube.

Water was collected and measured after flowing through the tube to calculate its concentration and calibrate the amount of injected water. A water content of 10 vol.% was used for these experiments. For testing in air, the alumina furnace tubes were not sealed. Up to 40 specimens were positioned in alumina boats in the furnace hot zone so as to expose the specimen faces to the flowing gas. After oxidation, selected specimens were Cu-plated and sectioned for metallographic analysis and electron probe microanalysis to determine Cr depletion.

TABLE 6-1 Alloy Chemical Compositions (weight %) and Average Grain Sizes (microns [μm]) of the 100-Micron Foil Materials

	Cr	*Ni*	*Mn*	*Si*	*Other*	*Grain Size (μm)*
Type 347	17.8	9.9	1.6	0.5	0.5Nb	5
20/25/Nb	20.3	24.7	1.0	0.4	1.5Mo, 0.2Nb	16
120	24.7	37.6	0.7	0.2		23
625	23.1	63.8	0.04	0.2	8.9Mo, 4Nb, 3Fe	12
Fe − 20/20 + MS	19.8	19.8	1.7	0.3	0.01Ce	25
Fe − 20/20 + 4Mn	20.9	20.8	3.8	0.2	0.3Nb, 0.3Cu, 0.3Mo	10
Fe − 15/15 + Al	15.1	15.8	4.8	0.2	3.8Al, 4Cu, 0.4Nb	30

(Source: Pint B. A. and K. L. More. 2004. Stainless steels with improved oxidation resistance for recuperators. *Proceedings of ASME Turbo Expo 2004*. American Society of Mechanical Engineers. Vienna, Austria.)

Results

Overview of Oxidation in Water Vapor

The effect of water vapor currently is being studied by several research groups, but there is no widely accepted mechanism for its role in reducing the corrosion resistance of chromia-forming stainless steels. Figure 6-4 gives a schematic representation of the current understanding of the role of water vapor on the oxidation of an austenitic stainless steel like type 347. In dry laboratory air, a protective Cr-rich oxide scale forms on the surface of type 347 stainless steel and thickens with time following a parabolic rate law.

Foil specimens (100 microns thick) have been exposed for more than 40,000 hours at 650°C and 25,000 hours at 700°C to confirm the long-term behavior. The addition of water vapor to the gas stream leads to effects such as faster oxide scale growth and increased evaporation of Cr_2O_3 from the scale as $CrO_2(OH)_2$. The net specimen mass change ($\Delta M_{specimen}$) measured after exposure in this type of environment can then be simply expressed as:

$$\Delta M_{specimen} = \Delta M_{oxide\ growth} - \Delta M_{evaporation} - \Delta M_{spallation}$$

Thus, if the loss due to evaporation is high enough, a net mass loss is measured without any oxide scale spallation. Both mechanisms result in an increased Cr consumption rate in the metal compared with oxidation in dry air. Because the diffusion rate of Cr in the metal is not fast enough relative to the consumption rate, a Cr-depleted region forms in the metal near the surface. After some incubation time, nodules of FeO_x begin to form.

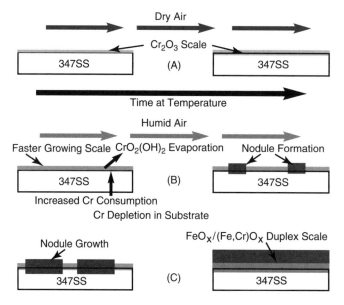

FIGURE 6-4   *Schematic of oxidation of type 347 stainless steel in dry air (A) and humid air (B,C). With minimal water vapor in the environment (a), a protective scale forms, which thickens with time at temperature. With the addition of water vapor (B), there is an increase in the scale growth rate and increased evaporation of Cr_2O_3 as $CrO_2(OH)_2$. These result in increased Cr consumption and depletion of Cr in the substrate near the surface. With continued exposure, nodules of FeO_x begin to form and grow with time (C). Eventually the nodules grow together, forming a duplex scale with an outer FeO_x layer and an inner $(Fe,Cr)O_x$ layer. (Source: Pint, B. A. and More, K. L. 2004. Stainless steels with improved oxidation resistance for recuperators. Proceedings of ASME Turbo Expo 2004. American Society of Mechanical Engineers. Vienna, Austria.)*

Whether the nodules form due to the Cr depletion in the adjacent metal substrate or some other mechanism has not been determined. However, it has been demonstrated that increasing the Cr and/or Ni content in the alloy increases the time to accelerated attack or prevents it from occurring. After the nodules form, they tend to grow laterally (Figure 6-4c) until a complete layer of Fe-rich oxide forms.

This results in a large specimen mass gain. An underlying layer of $(Fe,Cr)O_x$ also is observed. When these layers thicken sufficiently, they tend to spall during thermal cycling ($\Delta M_{spallation}$), which can lead to large specimen mass losses.

Oxidation Results at 800°C

Testing also is being conducted at higher temperatures to accelerate the corrosion testing and because higher-temperature recuperator materials may be needed. This temperature is beyond the capability of type 347 stainless steel from a creep-strength standpoint and accelerated attack was observed for all 347 foils in less than 1000 hours (Figure 6-5).

Several 20/25/Nb foil specimens have been run with varying results. In one case, the mass gain remained relatively low while in the other cases discrete jumps in the mass gain were observed followed by low mass gains or slight mass losses. These rapid increases appear to be localized nodule formation where the nodules did not continue to grow.

One foil specimen was removed from the test after 5000 hours whereas two others have been run to 6000 and 7000 hours, respectively, where they began to exhibit a continuous increase in mass gain, suggesting the onset of accelerated attack. Figure 6-6a shows the 20/25/Nb foil specimen after 5000 hours. Some oxide nodules were observed, but most of the scale was uniform and 3 to 4 microns thick.

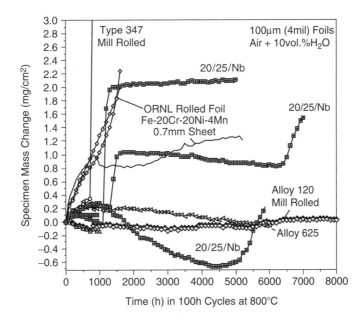

FIGURE 6-5 *Specimen mass gains for various foil (100 microns [μm] thick) materials during 100-hour cycles in humid air at 800°C. ORNL = Oak Ridge National Laboratory. (Source: Pint, B. A. and More, K. L. 2004. Stainless steels with improved oxidation resistance for recuperators.* Proceedings of ASME Turbo Expo 2004. *American Society of Mechanical Engineers. Vienna, Austria.)*

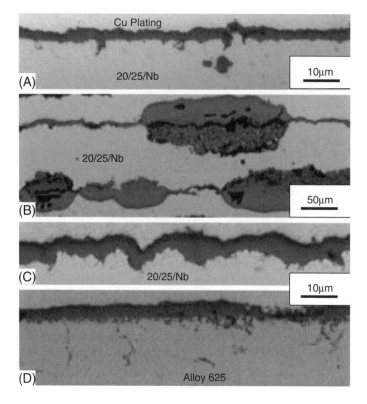

FIGURE 6-6 *Light microscopy of polished cross-sections of ORNL (Oak Ridge National Laboratory)-rolled 100-micron [μm] foils after exposure in humid air at 800°C: (A) 20/25/Nb for 5000 hours, (B, C) 20/25/Nb for 6000 h and (D) alloy 625 for 6000 hours. (Source: Pint, B. A. and More, K. L. 2004. Stainless steels with improved oxidation resistance for recuperators.* Proceedings of ASME Turbo Expo 2004. *American Society of Mechanical Engineers. Vienna, Austria.)*

Figures 6-6b and c show the 20/25/Nb specimens exposed for 6000 hours. In some areas, large oxide nodules were observed and, based on the mass gain increase over the last 1500 hours of exposure, these nodules were growing.

In areas without nodules, the scale was thicker and more convoluted than after 5000 hours (see Figure 6-6c). Despite the relatively low mass gain for this specimen, the formation of a relatively thick scale suggests that significant mass losses occurred due to evaporation (see equation above). The mass increase for an approximately 6-micron-thick scale should be 1 milligram per square centimeter.

Foil specimens of alloy 625 and 120 have not exhibited any rapid mass increases during testing at 800°C. Instead, they have shown relatively low mass gains or slight mass losses typical of the combination of scale growth and evaporation. The foil specimen of alloy 120 has been tested past 7500 hours without showing any signs of accelerated attack as was observed for 20/25/Nb. The longer time before accelerated attack for this alloy was expected because of its higher Cr and Ni contents.

Exposure of the alloy 625 specimen was stopped after 6000 hours for characterization. A cross-section of this foil is shown in Figure 6-6d. Again, a substantial oxide scale has formed, suggesting that significant loss of Cr occurred to result in a net mass loss after 6000 hours (see Figure 6-5).

Sheet specimens of the developmental Fe-20Cr-20Ni-4Mn alloy also are being tested (see Figure 6-5). The sheet specimen had a higher rate of mass gain during the first 1000 hours than the foil materials, but at longer times the rate of increase has decreased. Foil specimens of this material have only been tested for 2000 hours at this time and have shown a relatively high mass gain rate.

Testing sheet specimens of model alloys at 800°C has helped differentiate their performance (Figure 6-7). The specimen with Fe-20Cr20Ni has shown a longer time to accelerated attack than any of the other materials with lower Cr or Ni contents and Mn and Si additions.

Alumina scales are known to be more resistant to humid environments because Al_2O_3 is less susceptible to hydroxide formation. This has been shown for both alumina-forming alloys and aluminide coatings.

One alloy has been made with a 3.8% Al addition and has shown better resistance to accelerated attack than the other alloys. Because the corrosion resistance is due to the Al, the Cr and Ni contents were lowered to 15% to 16% (see Table 6-1).

Summary

The accelerated corrosion attack associated with the presence of water vapor in exhaust gas limits the temperature at which type 347 stainless steel can be used for a micro-turbine recuperator. Foil specimens of several candidate alloys are being studied in a long-term testing program.

These alloys generally show better corrosion resistance in these environments at 650°C to 800°C. A representative Ni-base alloy, 625, shows excellent corrosion

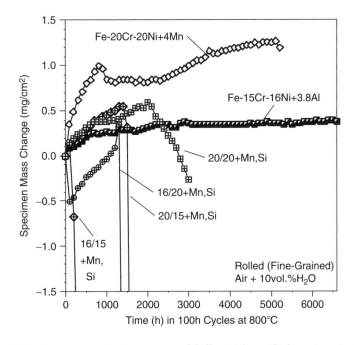

FIGURE 6-7 *Specimen mass gains for various model alloys (1.2-mm-thick specimens) with different Cr/Ni contents and additions of Mn and Si (labeled MS) during 100-hour cycles in humid air at 800°C. Adding 3.8% Al significantly improved resistance to accelerated attack. (Source: Pint, B. A. and More, K. L. 2004. Stainless steels with improved oxidation resistance for recuperators. Proceedings of ASME Turbo Expo 2004. American Society of Mechanical Engineers. Vienna, Austria.)*

resistance in foil form as did alloy 120 (Fe-37Ni-25Cr). However, these materials may be too expensive for microturbines.

Therefore, less expensive alternatives based on Fe-20Cr-20Ni and Fe-20Cr-25Ni are being explored in the ORNL alloy development program. Initial results show excellent corrosion resistance for these materials in humid air. However, the Cr depletion rates at 800°C suggest that these materials will be limited to approximately 700°C to 725°C if 40,000-hour durability is required. Alloys in this composition range are being selected for mill scale fabrication and eventual engine testing.

CASE 6-3***

Case 6-3 provides more details than Case 6-2 on recuperator alloy types and compositions, as well as their market readiness and current costs.

The DOE has an Advanced Microturbine Program with the goal to design and build microturbines with efficiencies of 40% or more. Recuperators with upgraded temperature capability and performance are an enabling technology toward this goal. While there are various types of compact recuperators, the main types used on commercial microturbines today are the primary surface welded air cells (Solar Turbines, Inc., design) in an annular recuperator configuration used by Capstone Turbines, Inc., and the brazed plate and fin (BPF) air cells in a stack recuperator configuration used by Ingersoll Rand Energy Systems.

ORNL has been conducting materials research and development in support of the Advanced Microturbine Program for several years to select, characterize, and develop materials with improved high-temperature performance for recuperators. Over the last year, ORNL has focused its efforts on:

(1) characterizing the properties of current commercial 347 steel sheet and foil used to manufacture recuperator air cells;
(2) identifying advanced alloys and/or modified commercial processing for making recuperators with upgraded performance;
(3) developing laboratory-scale modified 347 steels that offer the most cost-effective improvements in performance and reliability relative to standard 347 steel; and
(4) understanding the life-limiting mechanisms of actual engine-tested 347 steel recuperators.

These same steels and alloys are also being tested in the ORNL Advanced Microturbine Test Facility, which is based on a modified Capstone 60-kW microturbine.

Summary of Creep Resistance of Commercial Heat-Resistant Alloy Foils and Sheet

Previous initial screening work on a wide range of commercial or developmental heat- and corrosion-resistant alloys, processed as laboratory-scale foils at ORNL and creep tested at 750°C and 100 megapascals (MPa), established that HR120, and 625 alloys were the most cost-effective, high-performance recuperator alloy alternatives to 347 stainless steel. HR214 was identified as a possible higher-cost material for recuperator service above 800°C.

*** Adapted extracts from: Maziasz P. J., Pint, B. A., Shingledecker, J. P., Moore, K. L., Evans, N. D., and Lara-Curzio, E. 2004. Austenitic stainless steels and alloys with improved high-temperature performance for advanced microturbine recuperators. *Proceedings of ASME Turbo Expo 2004*. American Society of Mechanical Engineers. Vienna, Austria.

TABLE 6-2   Compositions of Heat-Resistant Austenitic Stainless Alloys Processed into Foils at Oak Ridge National Laboratory (weight %)

Alloy/vendor	Fe	Cr	Ni	Mo	Nb	C	Si	Ti	Al	Mn	Others
				Commercial stainless steels, alloys, and superalloys						
T347 steel	69.5	17.8	9.4	0.25	0.63	0.042	0.71	-	-	1.54	0.13 Co
NF 709	51	20.5	25	1.5	0.26	0.067	0.4	0.1	-	1.03	0.16 N
HR 120	39	25	32.3	1	0.7	0.05	0.6	0.1	0.1	1	0.3 Co, 0.2 N, 3 W max
HR 214	3.0	16	76.5	-	-	-	-	-	4.5		+ minor Y
alloy 625 (Special Metals)	3.2	22.2	61.2	9.1	3.6	0.02	0.2	0.23	0.16	-	
				ORNL developmental stainless steels						
Mod. 347-2	58.7	19.3	12.6	0.25	0.37	0.029	0.36	-	0.01	4.5	0.25 N, 4 Cu
Mod. 347-4	61.2	19.3	12.5	0.25	0.38	0.03	0.38	-	0.01	1.8	0.14 N, 4 Cu
Mod. 20/20	52.7	20.9	20.2	0.3	0.25	0.09	0.25	-	0.01	4.8	0.17 N, 0.3 Cu

Several demanding military recuperator applications employ alloy 625 for both primary surface and BPF air cells. Commercial alloy compositions are provided in Table 6-2. Experimental details regarding ORNL laboratory-scale foil processing to make 0.1-mm-thick foils, microcharacterization, and creep-testing have been given elsewhere.

Commercial 0.09-mm foil of HR120 alloy was obtained from Elgiloy Specialty Metals and tested for comparison to various foils and sheets of standard 347 steel used for commercial recuperator manufacturing. The modified alloy 803 was part of a small-scale development project between ORNL and Special Metals, Inc., but that alloy is not available commercially, so it is not included in this study.

Boiler tubing of NF709 stainless steel (Fe-20Cr25Ni,Nb,N; Table 6-3) from Nippon Steel Corp. was split, flattened, and then processed into foil for corrosion studies. In plate or tube form, NF709 is one of the most creep-resistant austenitic stainless steels at 700°C to 800°C. Therefore, it was also included in these most recent ORNL creep studies of foils for advanced recuperators.

New Processing for Improved Creep Resistance of Standard, Commercial Type 347 Stainless Steel Sheets and Foils

The recent ORNL program for upgrading the performance of commercial recuperator components began with establishing baseline creep behavior for the various commercial 347 stainless steel foils and sheets used by original equipment manufacturers for manufacturing recuperators. That initial work showed significant variability in the creep-rupture lives of standard, commercial foils and sheets of standard 347 steel (from 0.076 to 0.254 mm in gauge thickness) at 700°C to 750°C, ranging from 50 to 500 hours at 704°C and 152 MPa, with rupture ductilities from 3% to 27%.

These results clearly established a need for adjusting the processing for consistently better properties, so that a joint project between ORNL and the Allegheny Ludlum Technical Center (C. Stinner, PI) was established to produce commercial-scale quantities of 0.076-, 0.1-, and 0.127-mm foils and 0.254-mm sheet (the most common products used to manufacture primary surface or BPF recuperator air cells), with the processing parameters adjusted for improved creep resistance. The steel with more creep-resistant processing is now designated AL347HP™ and is commercially available from Allegheny Ludlum.

TABLE 6-3   Calculations of Relative Austenite Phase Stability Based on Alloy Composition

Alloy	Heat	Ni^A_{eq}	Cr^A_{eq}	Ni^B_{eq}	Cr^B_{eq}	Nibal	δ (%)
			Ni-Cr Equivalents for Austenitics				
Std. 347		11.4	20.2	11.4	19.0	−2.90	**13**
NF709		31.4	24.1	32.2	22.7	12.92	**0**
HR120		39.6	30.0	40.3	27.3	14.84	**0**
MOD 2 (Mod. 347)	18115	23.2	21.1	23.2	20.3	7.27	**0**
MOD 4 (Mod. 347)	18116	19.0	21.2	18.5	20.3	2.48	**0**
MOD 20/20	18529	29.7	22.3	30.4	21.7	12.50	**0**

$N i^A_{eq} = N i + Co + 0.5\,Mn + 30C + 0.3Cu + 25\,N$
$Cr^A_{eq} = Cr + 2Si + 1.5Mo + 5V + 5.5Al + 1.75Nb + 1.5Ti + 0.75W$
$Ni^B_{eq} = Ni + 0.5Mn + 30(C + N)$
$Cr^B_{eq} = Cr + Mo + 1.5Si + 0.5Nb$
$Nibal = Ni^B_{eq} - 1.36Cr^B_{eq} + 11.6$
δ from Schaffler-type Diagram (ASME Section VIII - Div. 1)

This project concluded with commercial sizes and quantities of foil and sheet appropriate for manufacturing BPF recuperator air cells shipped to Ingersoll Rand Energy Systems, and of foil for manufacturing primary surface recuperator (PSR) air cells shipped to Capstone Turbines.

Although creep resistance is one fundamental measure of improved temperature capability of foils for recuperator applications, resistance to moisture-enhanced oxidation is another such fundamental measure, and the AL347HP is likely to have similar oxidation/corrosion resistance to the standard T347 because the nominal steel composition was not changed. The coarser grain size of the AL347HP may also affect the formation of the protective surface oxide scale, but such effects must be determined by additional, systematic testing.

Alloys 625 and HR120 are commercially available sheet and foil materials with significantly better oxidation and creep resistance than standard commercial 347 steel at 650°C to 750°C, for enhanced performance and temperature capability of recuperators at about 3.5 to 4 times the cost of 347 steel. The NF709 (or similar Fe-20Cr-25Ni-Nb,N steel) and new ORNL modified 347 steels (containing Mn, N, and Cu) also have the potential to be more cost-effective alloys with upgraded performance and temperature capability for such recuperator applications, but they are not yet commercially available.

CASE 6-4[∧]

Case 6-4 describes some of the work in a Chinese university study on a commercially available microturbine. The interesting aspects of this case include a discussion on a rotating disk regenerator (potentially better heat transfer potential than a stationary regenerator if substantial air flow leaks can be avoided) and a manually controlled start sequence that explores the various speed thresholds in a start cycle. The Chinese researchers were also testing commercially available software to explore its potential use in further tests.

[∧] Adapted extracts from: Chiang, H.-W. D., Wang, C.-H., and Hsu, C.-N. 2004. Performance testing of a microturbine generator set with twin rotating disk regenerators. *Proceedings of ASME Turbo Expo 2004*. American Society of Mechanical Engineers. Vienna, Austria.

TABLE 6-4   Microturbine Generator Specifications

Manufacturer		Capstone	IR	Honeywell	Bowman	Turbec	Teledyne
Model		MD330	Powerworks	Parallon 75	TG80	T100	RGT-3600
Rated Power Output (kW)		30	70	75	80	100	150
Configuration (Shaft)		Single	Twin	Single	Single	Single	Twin
Shaft Seal		Air	Oil	Air	Oil	Oil	Oil
Net Electrical Effi-ciency (%)	With Heat-exchanger	27	33	27	26	30	28
	Without Heat-exchanger	14	-	-	14	-	-
Overall System Effi-ciency (%)	With Heat-exchanger	70	80	-	75	80	-
	Without Heat-exchanger	-	-	-	87	-	-
Engine Speed (rpm)		96,000	60,000	75,000	68,000	70,000	36,000
Heat Rate (L ~ (kJ/ kWh)	With Heat-exchanger	13,300	-	-	14,600	-	-
	Without Heat-exchanger	25,300	-	-	28,200	-	-
NOx Emission (ppm, O_2 15%)		<9	<9	<50	<25	<15	-
Noise Level (dBA)		65 10 m	-	65 10 m	77 1 m	70 1 m	-
Exhaust Gas Temp. (°C)	With Heat-exchanger	260	200	250	260	55	274
	Without Heat-exchanger	518	-	-	650	-	-

A 150-kW microturbine generator set with a Teledyne RGT-3600 microturbine was used in this study to investigate the basic steady-state performance of the generator set (Table 6-4). This investigation involved testing of the microturbine generator set at different load conditions using load banks. Using a personal computer–based data-acquisition system, the test data were recorded for different operating conditions. A software program was used to predict the performance of the microturbine generator set at different operating conditions to compare with the test results.

Microturbine Generator Set Features

The Teledyne RGT-3600 microturbine engine consists of a gasifier assembly, a power turbine, a combustor, a regenerator system, a reduction and accessory drive gearbox, and a fuel-management system, as shown in Figure 6-8.

The gasifier assembly consists of a single-stage, cast-aluminum compressor impeller at the front and a single-stage, axial flow gasifier turbine wheel attached to a

(A) Compressor and Diffuser   (B) Combustor

(C) Microturbine Engine

(D) Gasifier Turbine   (E) Ceramic Regenator Disk

FIGURE 6-8   *Major components of the Teledyne RGT-3600 microturbine engine. (Source: Chiang, H.-W. D., Wang, C.-H., and Hsu, C.-N. 2004. Performance testing of a microturbine generator set with twin rotating disk regenerators.* Proceedings of ASME Turbo Expo 2004. *American Society of Mechanical Engineers. Vienna, Austria.)*

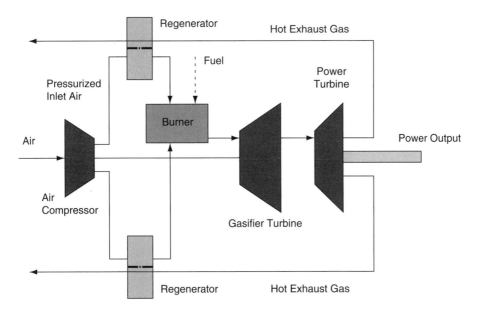

FIGURE 6-9 *Microturbine operation principle. (Source: Chiang, H.-W. D., Wang, C.-H., and Hsu, C.-N. 2004. Performance testing of a microturbine generator set with twin rotating disk regenerators.* Proceedings of ASME Turbo Expo 2004. *American Society of Mechanical Engineers. Vienna, Austria.)*

common shaft. A single can-type combustor mounts to the top of the microturbine with a single fuel nozzle and igniter. The regenerator system consists of two ceramic disks, disk seals, cast-iron regenerator covers, and a disk drive system.

As shown in Figure 6-9, air enters the single-stage radial compressor and discharge from the impeller flows through a vane-type diffuser. The air is directed to the regenerator disks at the sides of the engine. The compressor discharge air then flows inward through the regenerator disks to the combustor. The gases from the combustor expand first through the gasifier turbine and then through the power turbine.

The power turbine has variable nozzle guide vanes with different setting angles representing different engine operating lines. After expansion through the power turbine, the gases are diffused and directed outward through the two regenerator disks. The gases then exit from the regenerator covers into the exhaust pipes. Table 6-5 also lists the microturbine design performance.

Ceramic Rotating Disk Regenerators

For microturbines, the single improvement that will increase the thermal efficiency is the addition of a rotating regenerator or a stationary recuperator. The rotating regenerator is in general smaller and, for its size, is a much more efficient heat exchanger than the stationary recuperator. However, it does have the problem of leakage between the hot and cold gas streams. Figure 6-10 shows the flow pattern for the RGT-3600 twin rotating ceramic disk regenerators. Also, in operation, the temperature distribution in the regenerator produces a hot face and a cold face. For metallic regenerators, thermal expansion can cause the metal disk to bow into a spherical shape, with the hot face convex and the cold face concave, and creates a sealing problem.

TABLE 6-5   Microturbine Design Performance

Performance Parameters	Measured Data
Gasifier Speed	40,200 RPM
Power Turbine Speed	26,400 RPM
Output Shaft Speed	3000 RPM
Regenerator Speed	14.5 RPM
Max Rated Power	280 kW (375 hp)
Max Fuel Consumption	102 L/hr (27.0 gal/hr)
Pressure Ratio	4.1
Air Flow Rate	2.1 kg/sec (4.7 lb$_m$/sec)
Max Combustor Exit Temperature	1035°C (1895°F)
Exhaust Gas Temperature (EGT)	274°C (525°F)
Compressor Isentropic Efficiency	80%
Turbine Isentropic Efficiency	88%
Regenerator Effectiveness	89%
Engine Weight	816 kg (1800 lb$_m$)

The ceramic disk regenerator, on the other hand, has virtually zero thermal expansion and very much simplifies the sealing problem. The disk regenerator can be rotated by applying torque either at its rim or at its center. The RGT-3600 microturbine has a rim drive of the disk, which requires the incorporation of a gear drive system at the rim. The regenerator rotates at a design speed of 14.5 RPM.

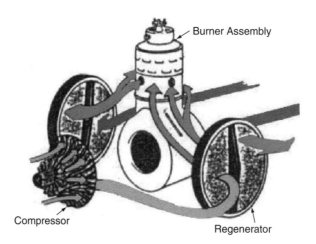

FIGURE 6-10   *Flow pattern for twin rotating disk regenerators. (Source: Chiang, H.-W. D., Wang, C.-H., and Hsu, C.-N. 2004. Performance testing of a microturbine generator set with twin rotating disk regenerators.* Proceedings of ASME Turbo Expo 2004. American Society of Mechanical Engineers. Vienna, Austria. Taken from: Detroit Diesel Allison, Division of General Motors Corp. 1977. *Allison gas turbines model GT404/505/606-310hp, 400hp, and 475hp industrial engines for military applications.* Division of General Motors Corp., SA 1658.)

FIGURE 6-11   *Microturbine engine test stand. (Source: Chiang, H.-W. D., Wang, C.-H., and Hsu, C.-N. 2004. Performance testing of a microturbine generator set with twin rotating disk regenerators.* Proceedings of ASME Turbo Expo 2004. *American Society of Mechanical Engineers. Vienna, Austria.)*

Engine Start Sequence

Because the microturbine engine did not come with the necessary start and control system including electronic engine control unit, a start sequence was developed and a manual control system installed. An engine testing facility was established for this purpose, as shown in Figure 6-11, which is composed of the test engine, a test stand, the instrumentation on the engine, air intake, and exhaust pipes. Without any factory manuals and guidance, it was rather complex to set up a working start sequence. A start sequence was finally developed for the microturbine engine as follows:

(1) Power on: Fuel boost pump on, auxiliary air pump on, and starter ready.
(2) Engine start command: Energize starter and start cranking.
(3) Gasifier turbine reaches light-on speed: Energize igniter and set starting fuel.
(4) Exhaust gas temperature reaches starting set point: Engine light on.
(5) Gasifier turbine speed reaches starting set point: De-energize starter and igniter, ramp up fuel to idle setting.
(6) Gasifier turbine speed reaches idle: Start sequence complete.

In addition, the gasifier turbine inlet temperature and speed, power turbine speed, variable power turbine nozzle guide vane schedule, lubrication, and exhaust gas temperature all need to be monitored during start, as shown in Figure 6-12. Figure 6-13 demonstrates a successful engine start sequence with a manual control system installed.

Microturbine Generator Set Testing

A microturbine generator set testing facility was established for this study, which is composed of the test engine, a three-phase AC 150-kW generator, the instrumentation on the microturbine engine, a load bank to simulate the loads, and a personal computer–based data acquisition system, as shown in Figure 6-14. The microturbine engine

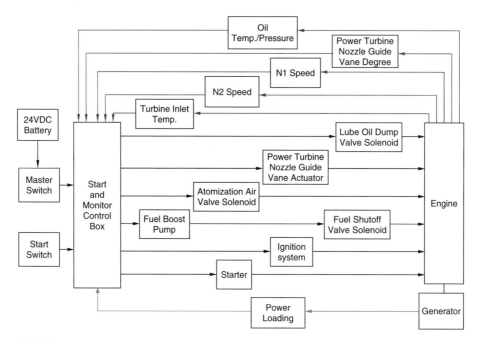

FIGURE 6-12 *Start and monitor system schematic. (Source: Chiang, H.-W. D., Wang, C.-H., and Hsu, C.-N. 2004. Performance testing of a microturbine generator set with twin rotating disk regenerators.* Proceedings of ASME Turbo Expo 2004. *American Society of Mechanical Engineers. Vienna, Austria.)*

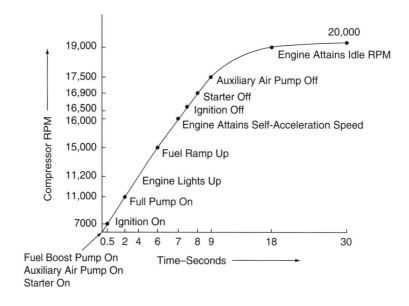

FIGURE 6-13 *Schematic of engine start sequence. (Source: Chiang, H.-W. D., Wang, C.-H., and Hsu, C.-N. 2004. Performance testing of a microturbine generator set with twin rotating disk regenerators.* Proceedings of ASME Turbo Expo 2004. *American Society of Mechanical Engineers. Vienna, Austria.)*

(A) Microturbine Generator Set Test Stand   (C) Load Bank

(B) Generator

FIGURE 6-14 *Instrumented microturbine generator set. (Source: Chiang, H.-W. D., Wang, C.-H., and Hsu, C.-N. 2004. Performance testing of a microturbine generator set with twin rotating disk regenerators.* Proceedings of ASME Turbo Expo 2004. *American Society of Mechanical Engineers. Vienna, Austria.)*

was supplied with sensors to measure temperature, pressure, fuel flow rate, and speed. The generator set was modified to mount the sensors and to provide instrumentation connections to the data-acquisition system.

Summary

An investigation was conducted to study the performance of a 150-kW microturbine generator set with twin rotating disk regenerators. This investigation involved testing of the microturbine generator set at different load conditions using load banks. Using a personal computer–based data-acquisition system, the test data were recorded for different operating conditions. A software program (GASTURB) was used to predict the performance of the microturbine generator set at different operating conditions to compare with the test results.

This investigation of the microturbine generator set has provided detailed insight of a microturbine application. The GASTURB software program predicted a thermal efficiency of 28% at full load with regeneration but only 14% with no regeneration. A good agreement between efficiency predictions and test data was found in the 0- to 100-kW actual test range, as limited by hardware problems.

Chapter 7

Microturbine Installation and Commissioning*

Installation and commissioning procedures are always specific to the make and model of microturbine in question. Installation and commissioning personnel are generally either original equipment manufacturer (OEM) personnel or the end user's staff who have been trained and certified by the OEM.

The basic tasks and precautions that are relevant to the installation and commissioning procedures are listed below. **For illustrative purposes, specific reference to installation and commissioning of the Elliott TA-100 model is made. Note all parameters and measurements as specific to this model.**

The installation and commissioning process for microturbines will generally consist of the following tasks:

1. OEM-provided training in all facets of the equipment in question
2. Training in model-specific and general safety procedures and hazards (e.g., electricity, fuel, hot surfaces, explosion)
3. Study of all the equipment-specific components (Figure 7-1), specifications (Figure 7-2), controls (Figure 7-3), safety systems, on-off switches, and electronics.
4. With installation,* consider all relevant site-selection factors, floor-planning requirements, and recommended service clearances.

SITE SELECTION*

Good installation planning is the key to proper site selection. Inadequate site planning may lead to future problems or potentially adverse operating characteristics for the microturbine. The following guidelines should be followed when selecting a site:

- Any potential site for the TA-100 must be free of debris, must be dry, and must not be subject to flooding.
- Exhaust gases must be properly ducted away from potential recirculation into the microturbine or away from any exposure to people or animals.
- When located indoors, the exhaust gases always need to be ducted to the outside.
- Distance between the TA-100 and any wall or permanent structure should be great enough to allow maintenance and easy removal of components. This distance, in the front (control panel end) and on each side of the TA-100, should not be less than 1.1 meters. The rear (exhaust end) of the turbine must have 1.5 meters of clearance.

---

* Excerpts taken from: Elliott Energy Systems, Inc. 2004. *Installation and commissioning manual EESI # 30.* Elliott Energy Systems, Inc. Stuart, FL.

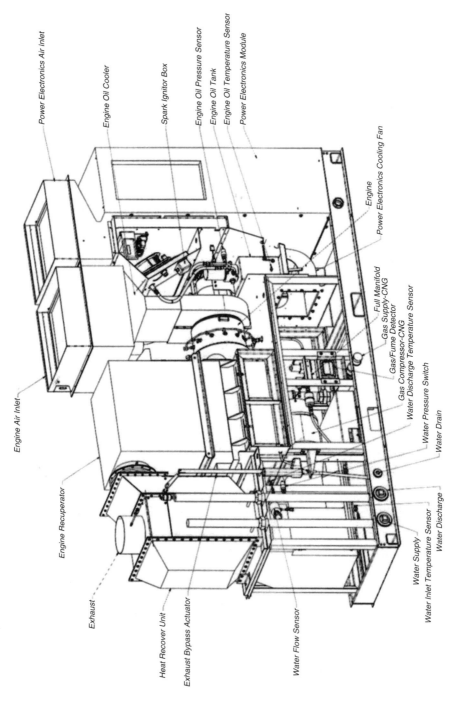

FIGURE 7-1 *TA-100 model microturbine combined heat and power major components. (Source: Elliott Energy Systems, Inc. 2004. Installation and commissioning manual EESI # 30. Elliott Energy Systems, Inc. Stuart, FL.)*

Performance:
Electrical

Output	100 kWe (+0/-3)
Turndown	100%
Efficiency	29% (+/-1) LHV

Fuel Consumption (ISO Rated Power)

CNG Recuperated:	0.62 m³/min
	362 kW (LHV)

Thermal Output (Hot water)

172 kW/587,000 Btu/hr.

Water Intel Temp	49°C
Water outlet Temp	60°C
Flow	4.55 L/s
Total System Efficiency	>75%

Engine Specifications

Manufacturer	Elliott Energy Systems
Model	TA-100
Type	Recuperated Gas Turbine
Pressure Ratio	4 to 1
Fuel Type	Natural Gas

Cooling System

Engine	Oil Cooled
Alternator	Oil Cooled
Inverter	Air Cooled
Enclosure Cooling	1.26 m³/min

Exhaust System

Outlet Size	25.4 cm
Rated Back Pressure	0 kPa
Max. Back Pressure	1.25 kPa

Fuel Supply

Pressure Required 0 - 0.345 Bar(G)

Lubrication System

Oil Type	Mobil SHC 824

Oil Capacity with Filter

19 L

Oil Filter	Spin On Type, 3 Micron

Emissions, Natural Gas

CO:
    <41 PPM @ 15% O_2
    <24 PPM Volume
    54.1 mg/MJ
    50 mg/m³ @ 15% O_2
    1.56 ibs/ MWhr
    0.50 grams hp
    0.13 ibs /MMBTU
    <24 PPM @ 15% O_2
    <14 PPM Volume
    51.9 mg/MJ
    50 mg/m³ @ 15% O_2
    1.49 ibs/ MWhr
    0.48 grams hp
    0.12 ibs /MMBTU

Exhaust Gas

Temperature	77°C
With Full Bypass	279°C

Batteries   12VDC min.
Battery Quantity: 2 (wired in Series),
YUASA NP 38-12, (38Ah / 12V
nominal voltage each.)

Total Weight with Enclosure:
1814 kgs

FIGURE 7-2 *TA-100 model microturbine combined heat and power specifications. LHV = lower heating value; ISO = International Organization for Standards; PPM = parts per million. (Source: Elliott Energy Systems, Inc. 2004.* Installation and commissioning manual EESI # 30. *Elliott Energy Systems, Inc. Stuart, FL.)*

- The site must allow adequate noise isolation from any people or surrounding facilities.
- The site must allow for the proper ducting to both the unit inlet air ducts and the engine enclosure cooling duct.
- The potential site must be within a reasonable distance of an adequate supply of fuel (natural gas).
- All natural gas pipes, water pipes, and electrical conduit routing must be done properly and in accordance with all applicable codes and regulations.

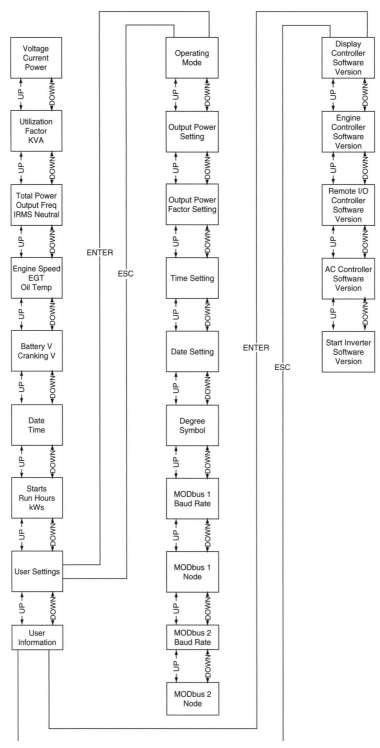

FIGURE 7-3   *Control panel menu structure. EGT = exhaust gas temperature; I/O = input/output; AC = alternating current. (Source: Elliott Energy Systems, Inc. 2004.* Installation and commissioning manual EESI # 30. *Elliott Energy Systems, Inc. Stuart, FL.)*

FLOOR PLANNING

- If installed outside, the TA-100 should be mounted and anchored to a 150-mm-thick concrete pad. The pad should extend at least 30 cm beyond the outside edges of the TA-100 cabinet. The concrete pad should be made of concrete rated at 21 kg/cm² and reinforced with fiber mesh.
- If installed inside or on top of a building, the TA-100 support pad must be strong enough to provide sufficient support. The TA-100 dry weight is 1814 kg.
- In all installations, the mounting surface must be level. Figures 7-4 and 7-5 show the proper tolerances for leveling the TA-100.

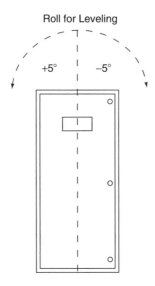

FIGURE 7-4 *Roll for leveling the TA-100 model microturbine. (Source: Elliott Energy Systems, Inc. 2004.* Installation and commissioning manual EESI # 30. *Elliott Energy Systems, Inc. Stuart, FL.)*

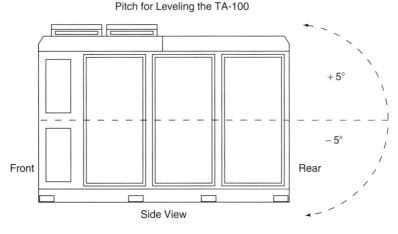

FIGURE 7-5 *Pitch for leveling TA-100 model microturbine. (Source: Elliott Energy Systems, Inc. 2004.* Installation and commissioning manual EESI # 30. *Elliott Energy Systems, Inc. Stuart, FL.)*

RECOMMENDED SERVICE CLEARANCES

The recommended service clearance for outdoor and indoor installations is the same, 1.1 meters minimum. This will allow for any maintenance to be performed on the units. Figure 7-6 shows an example of an outdoor installation, with given service clearances.

FACTORS AFFECTING PERFORMANCE

Altitude and Temperature Pressure Effects on Engine Inlet Air

Inlet pressure effects can be driven from two major sources. One source of pressure difference is induced by pressure losses due to long or restrictive air inlet ducts to the engine airflow system at the installation site. The other source of deviation is simply due to differences in altitude. Lower pressure due to altitude reduces the effective airflow available for work in the turbine.

Pressure losses in the inlet reduce the work output capability of the system since the turbine is a pressure ratio machine. With lower inlet pressure it cannot expand across the turbine as much as with a higher inlet pressure at the turbine. Figures 7-7 and 7-8 depict the overall system performance impact of both inlet pressure losses and altitude effects on the system.

The data contained in Figures 7-7 and 7-8 represent mean values that can be expected with the Elliott TA-100 combined heat and power (CHP) unit operating on natural gas. All data are subject to tolerances caused by manufacturing variations, installation differences, and measurement error. The tolerance for electrical power is +0, −3 kilowatts (kW) at maximum power at standard day reference at 1013 mbar 15°C. The tolerance of electrical efficiency is ±1% at maximum power under International Organization for Standards (ISO) conditions. The tolerance for thermal output is ±4 kW at the rated point. These data should be used as a reference in predicting the unit performance. Using these data points to provide financial guarantees is done at the guarantor's sole risk.

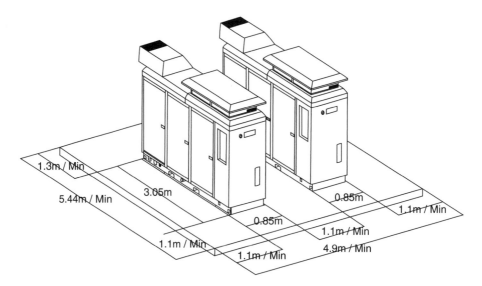

FIGURE 7-6   *Multi-unit outdoor installation. Min = minimum. (Source: Elliott Energy Systems, Inc. 2004.* Installation and commissioning manual EESI # 30. *Elliott Energy Systems, Inc. Stuart, FL.)*

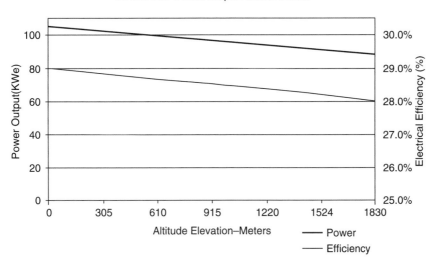

FIGURE 7-7 *Altitude derate. CHP = combined heat and power. (Source: Elliott Energy Systems, Inc. 2004.* Installation and commissioning manual EESI # 30. *Elliott Energy Systems, Inc. Stuart, FL.)*

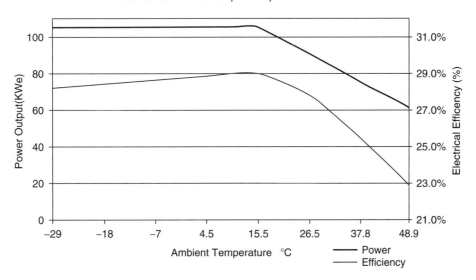

FIGURE 7-8 *Temperature derate. CHP = combined heat and power. (Source: Elliott Energy Systems, Inc. 2004.* Installation and commissioning manual EESI # 30. *Elliott Energy Systems, Inc. Stuart, FL.)*

SYSTEM AIRFLOW REQUIREMENTS

Engine Intake Airflow Requirements

The engine intake air is drawn into the TA-100 through a filter in a single 61 × 61–cm duct flange. The engine inlet system requires airflow at 755 L/sec nominal. See Tables 7-1 and 7-2 and Figure 7-9.

TABLE 7-1   Airflow Requirements

Inlet	Airflow	
Engine combustion	755 L/sec	45.3 m³/ min
Enclosure cooling	1085 L/sec	65.13 m³/min
Inverter cooling	354 L/sec	21.24 m³/min
Exhaust outlet	755 L/sec	45.3 m³/min

(Source: Elliott Energy Systems, Inc. 2004. *Installation and commissioning manual EESI # 30*. Elliott Energy Systems, Inc. Stuart, FL.)

TABLE 7-2   Maximum Back Pressure

Inlet	Maximum Allowable	Back Pressure
Engine combustion	12.7 mm H_2O	0.124 kPa
Enclosure cooling	12.7 mm H_2O	0.124 kPa
Inverter cooling	12.7 mm H_2O	0.124 kPa
Exhaust outlet	1.27 m H_2O	1.24 kPa

(Source: Elliott Energy Systems, Inc. 2004. *Installation and commissioning manual EESI # 30*. Elliott Energy Systems, Inc. Stuart, FL.)

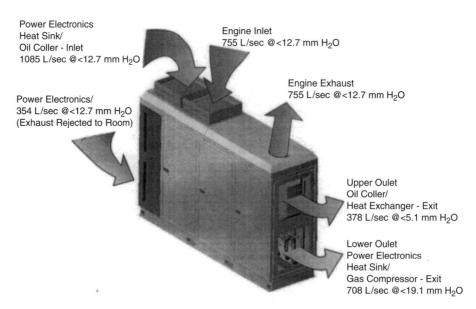

FIGURE 7-9 *Enclosure airflow requirements. (Source: Elliott Energy Systems, Inc. 2004.* Installation and commissioning manual EESI # 30. *Elliott Energy Systems, Inc. Stuart, FL.)*

Enclosure Cooling Flow Requirements

The enclosure cooling airflow is drawn into the TA-100 through a filter in a 61×61–cm duct flange and is driven by a 944-L/sec rated fan operating at 755 L/sec.

Oil Cooling Flow Requirements

The air-to-oil heat exchanger oil cooling flow is driven by a 472-L/sec rated fan operating at 401 L/sec.

Power Electronics Cooling

The TA-100 CHP is fitted with a spun-bonded glass-fiber filter that gives filtration to Eurovent EU3 or EN 779 Class G2 standards. The combined back-pressure for cooling air inlet and air outlet is 4.1 mm Aq (maximum) at 200 L/sec.

Cooling air flows through the power-conditioning unit. After passing over the electronics, the air exits the unit via the louvers in the top, left-hand side panel. Because the air is drawn in from, and is exhausted to the room, the room should be well ventilated to always ensure clean, cool air is available to cool the inverter.

INDOOR INSTALLATION

Air Inlet Ducting

Adequate airflow, both for the engine inlet air and for the enclosure cooling air, is critical for achieving overall peak performance. The engine intake and enclosure cooling airflow is drawn into the TA-100 through filters in two 61×61–cm duct flanges. One of these ducts provides engine combustion airflow, the other one provides cooling airflow for the enclosure. For indoor installations, ductwork should be fitted to both the engine intake and enclosure cooling flanges. The engine inlet air should be ducted from outside the building and should never come from inside an enclosed room.

Enclosure Cooling Exhaust Outlet

The enclosure cooling flow is exhausted from the unit through two openings, the upper and lower outlets, on the back of the unit enclosure.

The cooling exhaust from the upper and lower outlets must be ducted to the outside of the building to prevent preheating the inverter cooling air or the engine inlet air. The total air rejected from both the upper and lower outlets is 1085 L/sec.

Caution! The upper and lower enclosure outlets must be ducted separately for proper cooling of the enclosure. Failure to do so will not allow the enclosure to be properly cooled and could cause failure of equipment.

Engine Exhaust

The exhaust from the microturbine exits the top of the enclosure at the rear of the unit. Each application will have different exhaust requirements. The following guidelines should be followed when designing a system to duct the exhaust gases.

Exhaust ducting system should be designed as follows:

- Ensure that casual contact will not result in burns to personnel.
- The maximum allowable back-pressure is 1.27 m H_2O. The TA-100 is rated at ISO with no exhaust back-pressure. Therefore, any back-pressure will cause a slight derate in performance.
- Consider all safety issues associated with the high exhaust temperatures associated with the TA-100.
- Prevent rain or snow from accumulating in the exhaust pipe.

Exhaust technical data:

- Exhaust gas flow, 755 L/sec
- Diameter for exhaust pipe, 25.4 cm
- Maximum exhaust temperature, 302°C
- Maximum pressure drop from exhaust outlet to end of exhaust pipe should be less than 1.27 m H_2O.

Indoor Exhaust Connection

For indoor installations, the exhaust system must be designed to direct the microturbine exhaust gases outside of the building. The objective is to prevent burns and the accumulation of exhaust gases (Figure 7-10).

Outdoor Exhaust Connection

The outdoor exhaust manifold is designed to direct the exhaust up and away from the microturbine. This objective is to prevent burns and to prevent the exhaust from entering the engine air intake (Figure 7-11).

Note: If exhaust gases are drawn into the engine inlet air stream, the performance of the microturbine could be adversely affected.

Details to be noted include:

a. Fuel gas connections and check valve installation details (see OEM installation and commissioning manual).

FIGURE 7-10   *Indoor exhaust outlet. (Source: Elliott Energy Systems, Inc. 2004.* Installation and commissioning manual EESI # 30. *Elliott Energy Systems, Inc. Stuart, FL.)*

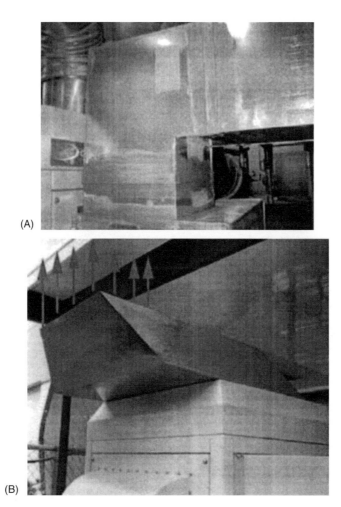

FIGURE 7-11   *(A) Enclosure outlet cooling duct. (B) Outdoor exhaust outlet. (Source: Elliott Energy Systems, Inc. 2004.* Installation and commissioning manual EESI # 30. *Elliott Energy Systems, Inc. Stuart, FL.)*

b. All power cable connections and grounding, protection relay, cable, and circuit breaker details (see OEM installation and commissioning manual)
c. Generator protection settings (Table 7-3).
d. Heat recovery unit connections, drains, bypass operation, valves, and performance ratings and limits (Table 7-4).

The following steps should be taken before commissioning is undertaken:

1. Complete all pre-commissioning checks with the safety and oil systems as specified.
2. Charge the system with oil as specified.
3. Conduct fuel and water leak checks as specified.

TABLE 7-3   TA-100 Model Microturbine Combined Heat and Power Generator Protection Settings

Parameter	Specification
Over-current (instantaneous)	300% of maximum rated current for two cycles
Thermal overload	Automatically ensures that the system will not supply more than maximum rated current by dropping terminal voltage. Under normal circumstances, if excessive load is placed on the generator, the system will shut down on under-voltage.
Ground leakage	100 mA for 60 milliseconds
Ground over-voltage relay	Operates if more than 5 V exists between chassis earth and grounded neutral.
Over-voltage	110% of nominal RMS rated voltage
Under-voltage (in voltage mode)	90% of nominal RMS rated voltage
Under-voltage (in utility mode)	90% of nominal RMS rated voltage
Under/Over-voltage time delay	833 milliseconds for 69 Hz; 1000 milliseconds for 50 Hz
Over-frequency	101% of nominal frequency
Under-frequency	96% of nominal frequency
Under /Over-frequency time delay	667 ms for 60 Hz; 800 ms for 50 Hz

RMS = root mean square.
(Source: Elliott Energy Systems, Inc. 2004. *Installation and commissioning manual EESI # 30*. Elliott Energy Systems, Inc. Stuart, FL.)

TABLE 7-4   TA-100 Model Microturbine Combined Heat and Power HRU Performance Limits

HRU Performance Limits		
Water Inlet	Water Outlet	Pressure Relief
50 mm	50 mm	25 mm
Maximum Rated Conditions		
10.3 bar g	10.3 bar g	10.3 bar g
303 L/min Max	303 L/min Max	303 L/min Max
82°C	93°C	93°C

(Source: Elliott Energy Systems, Inc. 2004. *Installation and commissioning manual EESI # 30*. Elliott Energy Systems, Inc. Stuart, FL.)

4. Check batteries and battery connections as specified.
5. Complete safety checklist.
6. Study all operating procedures specified, including different operating modes, stop and start p rocedures, commissioning check lists (Table 7-5), configuration and communication connections, alarms both remote and local, and shutdown sequences.
7. Proceed with commissioning according to the itemized steps provided in the OEM installation and commissioning manual.

TABLE 7-5   TA-100 Combined Heat and Power

Microturbine Commissioning Checklist

Date_____   Site Name _____

Customer Name_____

Serial Number(s) - Unit A _____ Unit B_____

Unit C _____ Unit D _____

Unit E _____

On-Site Contact_____ On-site Tel ____ _____ _____

Distributor Name _____ Distributor Tel: ____ _____ _____
(Please include country code of telephone numbers when outside the United States.)

Installation Address _____

City _____State _____ Postal Code _____

Country _____

Application Data - Frame Size _____ KW ▮ Unit Placement _____ ▮ Software Version _____

GenNumber_____ ▮ Engine S/N _____ ▮ Module S/N ____ ˉ____ ˉ____

Type of Cogeneration _____ ▮ Fuel   ▮ Site Elevation _____ FL _____ meters

Voltage _____ ▮ Frequency _____ ▮ Phase _____ ▮ 3 Wire   Or 4 Wire (4 Wire requires a transformer)

Transformer Information ▮ Type - Isolation   - Step Down   - Step Up   - 3 to 4 wire conversion   ▮ Output Voltage

Ratings_____ kW - _____ kva ▮ Wet   - Dry ▮ Efficiency___% Impedance _____Ohms

Manufacturer _____ ▮ Model _____

Safety Checklist - Follow the safety cheklist instructions in the "Installation and Commissioning Manual" EESI # 11. Check each of the boxes below when completed.
Electric   ▮ Fuel   ▮ Intake Filter   ▮ Exhaust   ▮ Gas Safety Shut Off Value Installed

Main Electrical Disconnect Installed

Pre-Start Inspection - Follow the safety cheklist instructions in the "Installation and Commissioning Manual" EESI # 11. Check each of the boxes below when completed.
Interconnectiion Checklist
Utility Connection   ▮ Fuel Connection   ▮ Water Inlet ▮ Water Outlet   ▮ Engine Inlet   ▮ Electronics Cooling Inlet

Exhaust Outlet   ▮ Communications   ▮ Oil Level   ▮ Air Filters Installed Fuses Engaged   ▮ Batteries Charged

Batteries Connected   ▮ Unit Grounded Properly   ▮ Chiller Unit Communications Interface Installed. (If Applicable)
Load Following Equipment Installed

Power On Tests Oil Pressure   PSI (Oil Pressure must be ____ +/- ____ PSI ▮ Battery Voltage   VDC

Using the Commissioning Tool software verify that the following are operating. See "Installation and Commissioning Manual" EESI # 11 for instructions on how to use the software. Oil Cooler Fan ▮ Igniter   ▮ Power Electronics Cooling Fan ▮ Gas Compressor

Enclosure Cooling Fan   ▮ Fuel Supply Stop Valve ▮ Verify: Accurator Stroke   ▮ Configuration Set to Proper Operation Mode

Operational Tests - Log all starts during the commissioning process using the Commissioning Tool software.

Output Voltage Correct   ▮ Output Frequency Correct   ▮ Output Power Correct

Operate at full load for 3 hours before finishing the commisssioning checklist.

Normal Operations Checks (Warning: Do not leave doors open longer than three minutes!! Open doors disrupt the flow of cool air throuth the electronics bay.)

Water Temperature in _____ °F/ _____°C   ▮   Water Temperature in _____ °F/ _____°C   ▮ Water Temperature Stablized

Water Flow _____ GPM / _____ LPM ▮ Water Temperature _____ GPM / _____ LPM

Remote Communication Working Properly   ▮   Power Output _____ kW ▮ Verify unit shuts down when gas safety shut off valve is closed.   ▮ Restart Unit

Post Commissioning Tasks Check for Oil Leaks   ▮ Check for Water Leaks   ▮ Check for Fuel Leaks

Remove all lools and installation materials from site   ▮ Make sure all tools are accounted for   ▮ Clean site

Review site with customer representative   ▮ Does installation meet the site specifications? Yes   No   (If no unit is not considered commissioned)   ▮ Give keys to building or facility manager of other responsible customer representative

(Continued)

TABLE 7-5 TA-100 Combined Heat and Power—Cont'd

Signature of Customer Observing the Commissioning　　　　————————————————————————

(By signing above you agree that the equipment has beeb installed to your satisfaction.)

Print Name ————————————————————　　Date ——————————

Signature of Technician Performing Commissioning　　　　————————————————————————

(By signing above you certify that all checks have been made and that the installation meets all the conditions for proper operation.)

Print Name ————————————————————　　Date ——————————

(Source: Elliott Energy Systems, Inc. 2004. *Installation and commissioning manual EESI # 30*. Elliott Energy Systems, Inc. Stuart, FL.)

Chapter 8

Microturbine Maintenance, Availability, and Life Cycle Usage

MAINTENANCE

A learning curve still exists in terms of the maintenance of microturbines, as most initial commercial units have seen only 2 to 4 years of service so far. With relatively few operating hours logged for any single unit, the population of microturbines in the field has not yet provided enough long-term operational data to allow for a clear definition of durability and maintenance costs.

Most manufacturers offer service contracts for maintenance priced at about $0.01 to $0.02 per kilowatt per hour. The combustor and associated hot section parts are inspected periodically. Air and oil filters are replaced periodically, and oil bearings are inspected. Microturbines operating in environments with extremely dusty air require more frequent air-filter changes.

A microturbine overhaul is needed every 20,000 to 40,000 hours, depending on manufacturer, fuel type, and operating environment. A typical overhaul consists of replacing the main shell with the compressor and turbine attached, general inspection, and, if necessary, replacing the combustor. During overhaul, other components are examined to determine whether wear has occurred, with replacements made as required.

Microturbines in peak-shaving service are usually operated with at least one on-off cycle per day. There are general concerns about the effects of this type of operation on component durability due to thermal cycling, which causes material fatigue. There are no established differences in maintenance for operation on fuels other than natural gas. However, experience with liquid fuels in industrial gas turbines suggests that liquid-fueled combustors should be inspected more frequently than natural gas-fueled combustors and may require more frequent maintenance. Microturbines that use still harsher fuels, such as biomass slurry, will tend to show still more wear, as compared with gas fired counterparts.

AVAILABILITY AND LIFE

With the small number of units in commercial service, information is not yet sufficient to draw conclusions to the same extent as one might with, for instance, gas turbines, about the reliability and availability of microturbines in different applications. The basic design and low number of moving parts promote the potential for systems of high availability. Manufacturers have targeted availabilities in the range of 98% to 99%. The initial projection of availability for early market microturbine systems is 95% and up. The use of multiple units or backup units at a site can further increase the availability of the overall facility.

The equipment life of microturbines is estimated to be 10 years; this includes at least one major overhaul in that time frame. Costs of these overhauls are included in the non-fuel related maintenance item estimates for calculating costs per fired hour. The economic life of microturbine systems is typically assumed to be 10 years.

Part 2

Microturbine System Applications and Case Studies

Chapter 9

Microturbines Operating in Power-Only Applications

This chapter deals with microturbines that operate on their own, without benefit of added power from a fuel cell, diesel genset, or other means of producing power. Microturbine outputs are typically 30 to 400 kilowatts, but there is no real "fixed" size bracket at this time. The field is a new and evolving one.

Various microturbine applications are discussed so the reader gets an idea about the scope of microturbines as a stand-alone item of equipment.

As is said in other places in this book, the microturbine has some marketing and image problems with being thought of as "a fashionable fad" by conventional large power consumers, who would rather their world did not change substantially in favor of distributed energy anytime soon. Conventional power generation people in the mainstream U.S. market are wary of anything that could stem from the "green team." The operative phrase here is *in the mainstream U.S. market*. In many areas of the United States that are untraditional in their outlook, such as the state of California, the idea of a small factory or school or hospital being energy independent, works better than it might in West Virginia. The "always have a backup just in case" mentality may be blamed for such microturbine users also retaining a connection to the local power grid. This happens in the rest of the highly industrialized world, too (see Case 9-8).

The microturbine industry is in its relative infancy and, therefore, so are absolutely reliable, top-notch service teams; established spare parts and component suppliers; and established life-cycle patterns for all major components. So the "backup" may in fact prove to be prudent for all microturbine applications, depending on the service factors in question.

However, in many areas in the rest of the world that have no grid, no infrastructure, and not enough money to get either, the possibility of a microturbine supplying independent power is a huge boon. They could even overlook some outages and service problems because that would be so much better than what it replaced.

In terms of greenhouse gas emissions, microturbine installations can help reduce the carbon dioxide load indirectly. When they are used "solo" and eliminate grid line connections, the forest that might otherwise have been cut down to allow laying those grid lines can continue to absorb carbon dioxide instead.

The real issue with microturbines, however, is that—for all the reasons cited above and in Chapter 15: Business Risk and Investment Considerations—they are not yet mainstream. *Mainstream* could be defined as enough business volume to produce a financially stable pool of manufacturers. That kind of volume will happen when and if individual consumers get involved and want their own "mini" microturbine to power their houses in *sufficiently large numbers*. U.S. "green thinking" individual consumers can buy a house-sized wind turbine or solar generator just as inexpensively as a small microturbine and may find the former sources more reliable, in terms of dependability and service.

In Europe, village co-ops in countries like Denmark are more likely than their U.S. counterparts to buy a 1.5-megawatt (MW) wind turbine and share it. The wind turbine, given its culture and history, is a familiar sight in Europe. A microturbine is not accepted as well as wind or solar power in fossil fuel-poor countries. So Europe is unlikely to be this "household-size" microturbine market base unless it also has a source of some microturbine fuel.

That notwithstanding, there will always be some schools, businesses, and hospitals for whom microturbines are the right choice. However, making them the total energy supply choice may not be practical (see Case 9-8). They may be some part of the right choice, as some case histories illustrate. The following case histories all occur in Japan and New Zealand, and were drawn from the commercial literature of original equipment manufacturers (OEMs) such as Mitsubishi and Capstone.

CASE 9-1: CHEMICAL PLANT[*]

At a very large chemical production facility in Japan, power from 44 C60s (OEM: Capstone) cuts power costs, saving about $500,000 per year. Exhaust from the array is ported directly into polymer-drying ovens, off setting gas purchases. This would not be possible with any other power generator. Since these microturbine systems use no fluids whatsoever, the dry exhaust stream is uncontaminated by oil or coolant vapor (which would otherwise render the polymer product useless).

CASE 9-2: FACTORY COMMISSARY[*]

At one of their factories in Japan, Mitsubishi has designed one of their own combined heat and power (CHP) hot water systems designed around a microturbine to provide power, air conditioning and desiccation, and hot water for an in-house commissary/ convenience store.

CASE 9-3: HOSPITAL[*]

At a small hospital in Japan, an array of eight kerosene-fueled Meidensha Miospectrum original equipment manufacturer CHP systems mitigate power and water-heating costs.

CASE 9-4: DAIRY FARM[*]

A microturbine uses fuel from a modern biogas treatment facility at a Japanese dairy farm. Solid and liquid waste from 1000 cows produces biogas that is routed through a digester system. The microturbine's hot exhaust is diverted back through the digester, which continues a closed loop cycle. Excess power is exported to the local utility grid.

CASE 9-5: MASS TRANSIT HEV BUSES[*]

A fleet of four microturbine-powered buses in New Zealand have accumulated more than a half-million miles worth of service in their 14-hour-per-day schedules.

---

[*] Reference: Capstone Turbines, CA website and other Capstone literature.

CASE 9-6: OFFICE/FACTORY BUILDING[*]

In an eight-story office building in Tokyo, a microturbine supplements building power needs. Hot water from the building's CHP system is routed to a nearby absorption chiller.

CASE 9-7: UNIVERSITY[*]

A Japanese university gets summer absorption chilling and winter heating from their microturbine and CHP system.

CASE 9-8: CONSUMER ELECTRONICS MANUFACTURING[*]

At a Japan factory of a consumer electronics manufacturer, a six-pack microturbine system provides the primary power source of choice to sensitive manufacturing and communications systems. Uninterrupted-power-supply batteries are primary backup, and the grid serves as the final source of emergency power.

Note that some of the above cases describe CHP applications that are also discussed in Chapter 10: Combined Heat and Power with Microturbines. The point the author is making here is that although *CHP* generally refers to power and waste heat absorbed in medium-sized or industrial applications, it makes good sense to always use waste heat from any turbine to heat water, air, or whatever else needs heating.

For basic engineering details on simple (stand-alone) microturbine systems, see Part 1 of this book and the Index. Tables 9-1 through 9-7 and Figures 9-1 through 9-4 provide a comparison of the cost of simple microturbine systems versus other small energy packages:

TABLE 9-1 Microturbines Versus Central Plant: Energy Comparison

Central Plant (Combined Cycle, Natural Gas, 50% Efficient)		*Microturbine in CHP (60 kW, 26% LHV, 75% Overall)*	
	kW/hr		**kW/hr**
Power delivered	57.9	Power delivered	57.9
Line loss (6%)	3.69	Parasitic	2.1
Power generated	61.59	Power generated	60
Fuel to generator	123.19	Fuel to generator	231
Building heat	117	Building heat	117
Fuel to boiler	148	Fuel to boiler	0
(Correction for LHV)	133.2		
Total energy input	256.39	Total energy input	231
		Input difference	25.39
		Fuel savings: 2.73 m³/hr, or 10%	

CHP = combined heat and power; LHV = lower heating value.
(Source: Whitehead T. 2003. *Clean energy opportunities: microturbine in cogeneration.* Enbridge Gas Distribution Environment and Energy Conference.)

TABLE 9-2   Microturbines in Cogeneration: Emissions Comparison of Technologies

	CO_2	NO_x
Microturbine	294 kg/MWhr[*]	−0.008 kg/MWhr[*]
Central plant (NG combined cycle)	350 kg/MWhr	0.25–0.5 kg/MWhr
Central plant (coal)	1000 kg/MWhr	2 kg/MWhr
Rich burn reciprocating engine (post emissions control)	291 kg/MWhr[*]	0.095 kg/MWhr[*]
Lean burn reciprocating engine	288 kg/MWhr[*]	0.88 kg/MWhr[*]
Fuel cell (35% efficient)	288 kg/MWhr[*]	−0.175 kg/MWhr[*]

NG = natural gas.
[*]Emissions corrected to reflect generator contribution of cogeneration system, based on 75% overall, and allowable boiler.
(Source: Whitehead T. 2003. *Clean energy opportunities: microturbine in cogeneration.* Enbridge Gas Distribution Environment and Energy Conference.)

TABLE 9-3   Cost of Emission Reduction for Microturbines Versus Other Methods

	Microturbine	*Lean Burn Reciprocating*	*Rich-burn Reciprocating with 3-way Catalyst*
Output	60 kW	60 kW	60 kW
Est. Installation Cost	$155,000	$135,000	$143,000
Annual emissions equipment maintenance[1]	0	0	$3000
Annual operating savings[2]	$20,000	$20,000	$20,000
Simple payback	7.75 years	6.75 years	8.4 years
NO_x reduction (coal)	1018 kg	568 kg	1062 kg
Incentive needed to reach 5-year simple payback	$1250/kW	916/kW	$1250/kW

[1]NREL/SR-560-31772, October 2002, "The Impact of Air Quality Regulations on Distributed Generation."
[2]Estimated savings based on 8760 hours of operation annually, average fuel costs $0.25/m³, $66/MWhr, 75% total efficiency.
(Source: Whitehead T. 2003. *Clean energy opportunities: microturbine in cogeneration.* Enbridge Gas Distribution Environment and Energy Conference.)

TABLE 9-4   Capacity, Efficiency, and Cost Data for Independent Power Generation Systems

Power System	*Capacity Range*	*Electrical Efficiency (HHV)*	*Capital Cost ($/kW)*	*Operating & Maintenance Cost ($/kWh)*
Reciprocating engine	20 kW to 20 MW	28% to 45%	$500 to $1400	$0.007 to $0.02
Microturbine	~25 to 300 kW	~20% to 33%	$600 to $1000	$0.003 to $0.01
Gas turbine	500 kW to 150 MW	21% to 40%	$600 to $900	$0.003 to $0.008
Fuel cell	5 kW to 3 MW	36% to 60%	$1900 to $3500	$0.005 to $0.10
Stirling engine	~200 W to 100 kW	20% to 36%	$1000	Not available
Rotary engine	~5 kW and up	20% to 30%	Not available	Not available

(Continued)

TABLE 9-4 Capacity, Efficiency, and Cost Data for Independent Power Generation Systems—Cont'd

Power System	Capacity Range	Electrical Efficiency (HHV)	Capital Cost ($/kW)	Operating & Maintenance Cost ($/kWh)
Photovoltaic	1 kW to 1 MW	6% to 19%	$6600	$0.001 to $0.004
Wind turbine	10 kW to 1 MW	25%	$1000	$0.01

HHV = higher heating value.
(Source: New Technology Demonstration Program, Federal Energy Management Program. 2000. *Integrated systems*. U.S. Department of Energy. Washington, D.C.)

TABLE 9-5 Advantages and Disadvantages of Technologies for Independent On-Site Power Generation

Generating Technology	Advantages	Disadvantages/Problems
Microturbine generators	High reliability	Require high-pressure gas or gas compressor
	Compact and modular design	Not able to start under large load or follow large transients
	Low maintenance and operating costs	Power electronics need further development
	Low emissions and noise	May be life cycle problems with recuperator
	Ease of operation	Performance sensitive to temperature and altitude
Reciprocating engine-generator sets	Run well on part loads	Emissions
	Suitable for start/stop operation	Noise
	Can be sized for lower electrical loads	Efficiency
	Follow electrical and thermal loads	
	Insensitive to temperature and altitude	
Combustion gas turbine generators	Reliable technology	Need to be run on constant load
	High efficiency	Not suitable for start/stop operation
	Low maintenance requirement	
	High-quality heat: 20,000–25,000 lb/h 125 psig steam (5-MW plant)	
Fuel cells	High efficiency	Poor ability for multiple starts
	High-output power quality	Poor ability to follow large, rapid transients
	No moving parts	High capital cost
	Very low emissions	Availability

(Continued)

TABLE 9-5   Advantages and Disadvantages of Technologies for Independent On-Site Power Generation—Cont'd

Generating Technology	Advantages	Disadvantages/Problems
	Low noise	Not a firmly established technology
Photovoltaic and wind generators	Pollution free	Dependent on availability of environmental resources

(Source: New Technology Demonstration Program, Federal Energy Management Program. 2000. *Integrated systems*. U.S. Department of Energy. Washington, D.C.)

TABLE 9-6   Approximate Sizing of Electrical Load and Building Sizes

Magnitude of Electrical Load	Type of Application
>1 MW	Large high-rise office buildings
	Largest hospitals
	Largest hotels
	Large shopping malls
200 kW to 1 MW	Hospitals (200 to 300 beds)
	Large hotels (750 rooms)
	Office buildings (200,000 ft^2)
	Schools (125,000 ft^2)
	Large retail buildings
50 to 200 kW	Office buildings (50,000 ft^2)
	Average hotel (75,000 ft^2, 125 rooms)
	Multi-family residences (100 units)
10 to 50 kW	Fast food restaurant (4000 ft^2)
	Small office building (10,000 ft^2)
	Multi-family residences (<25 units)
~10 kW peak load	Single-family residence
0.50 to 1.5 kW average load	
0.10 kW base load common	
Little coincidence of electrical and thermal loads	

(Source: New Technology Demonstration Program, Federal Energy Management Program. 2000. *Integrated systems*. U.S. Department of Energy. Washington, D.C.)

TABLE 9-7   Heat Recovery, Maintenance Schedule, and Emission Data for Generating Systems

Power System	Maximum Heat Recovery Temperature (°F)	Heat Recovery (BTU/h per kW)	Expected Time Between Overhaul (Operating Hours)	NO$_x$ Emissions (ppm)
Reciprocating engine	~200°F water jacket	~4000 to 10,000	25,000+	20

(Continued)

TABLE 9-7 Heat Recovery, Maintenance Schedule, and Emission Data for Generating Systems—Cont'd

Power System	Maximum Heat Recovery Temperature (°F)	Heat Recovery (BTU/h per kW)	Expected Time Between Overhaul (Operating Hours)	NO_x Emissions (ppm)
	750 °F to 930 °F exhaust			
Microturbine	~500 °F	~4000 to 12,000	40,000+	<1
Gas turbine	930 °F to 1100 °F	Not available	Not available	Not available
Fuel cell	140 °F to 180 °F (PEM)	~3500 to 4000	40,000+	~1
	390 °F (PAFC)			
	1100 °F to 1400 °F (SOFC)			
	~1100 °F (MCFxC)			
Stirling engine	~160 °F to 200 °F	~6000 to 12,000	up to ~60,000+	Not available
Rotary engine	~300 °F water jacket	Not available	Not available	Not available
	~1600 °F exhaust			

PEM = proton exchange membrane fuel cell; PAFC = phosphoric acid fuel cell; SOFC = solid oxide fuel cell; MCFC = molten carbonate fuel cell.
(Source: New Technology Demonstration Program, Federal Energy Management Program. 2000. *Integrated systems*. U.S. Department of Energy. Washington, D.C.)

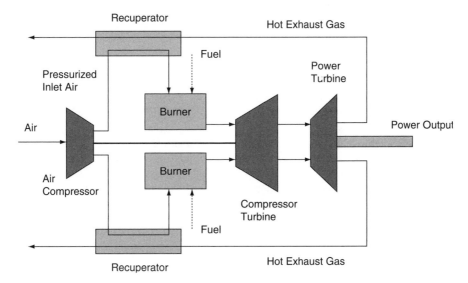

FIGURE 9-1   *Schematic of a microturbine generator. (Source: New Technology Demonstration Program, Federal Energy Management Program. 2000.* Integrated systems. *U.S. Department of Energy. Washington, D.C.)*

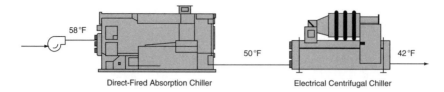

FIGURE 9-2   *Series layout of gas and electric chillers. (Source: New Technology Demonstration Program, Federal Energy Management Program. 2000.* Integrated systems. *U.S. Department of Energy. Washington, D.C.)*

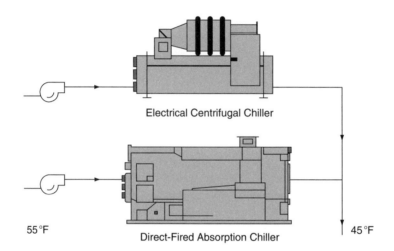

FIGURE 9-3   *Parallel layout of gas and electric chillers. (Source: New Technology Demonstration Program, Federal Energy Management Program. 2000.* Integrated systems. *U.S. Department of Energy. Washington, D.C.)*

PROJECT DEVELOPMENT[*]

Case Study

District heating does not necessarily have to produce both heat and power, but often this is the case. It is also not limited to just downtowns, but can also include "campus heating" of educational and other multiple-building facilities. St. Johns University near St. Cloud, MN, has a CHP campus heating system that uses a coal and waste wood fired steam unit. The Mayo Clinic in Rochester is also supplied by a CHP system.

Waste heat from local processing facilities also presents an opportunity for community-wide heating and cooling systems. This would both promote private-public cooperation and decrease the energy usage of the entire community.

The West Central Research and Outreach Center and the University of Minnesota–Morris are working with DENCO, a farmer-owned ethanol plant, to utilize the waste steam heat that DENCO would generate. The University of Minnesota–Morris would

---

[*] This section contains excerpts from: Pawlisch, M., Nelson, C., and Schoenrich, L. 2003. *Designing a clean energy future: developed for the clean energy resource teams*. Minnesota Department of Commerce. St. Paul, MN.

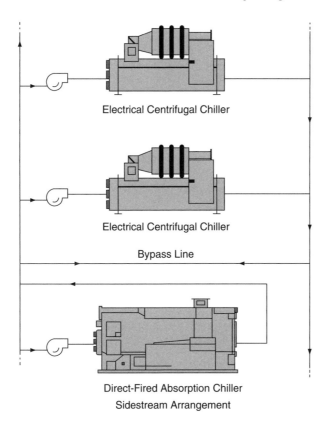

Electrical Centrifugal Chiller

Electrical Centrifugal Chiller

Bypass Line

Direct-Fired Absorption Chiller
Sidestream Arrangement

FIGURE 9-4 *Sidestream layout of gas and electric chillers. (Source: New Technology Demonstration Program, Federal Energy Management Program. 2000.* Integrated systems. *U.S. Department of Energy. Washington, D.C.)*

use this waste steam in a district energy system that would serve its needs and those of a new elementary school while allowing DENCO to recover some of its costs.

Installing district energy systems is not without obstacles. These systems require significant capital investment to create the necessary infrastructure support. This means that district energy systems need community support, but district energy presents a real solution for improved energy efficiency and presents a tangible way for communities to reduce their fuel consumption.

Case Study: Virginia Department of Public Utilities: CHP at a Local Utility

The Virginia Department of Public Utilities is located in Virginia, MN, along Minnesota's iron range. The current power plant operates a 30-MW CHP power plant that consists of three boilers and four turbines and burns primarily western coal and natural gas, depending on the boiler.

Electricity is produced by the power plant to fulfill the demands of the steam system. The steam district heating system supplies 2500 customers including the downtown business area, city public buildings, and south side and north side commercial and residential areas while the electric system serves over 5800 customers. Recent construction activities have forced the closing of steam lines to particular neighborhoods, reducing the number of homes served by steam heat. Overall, however, the CHP district heating system in Virginia has proven to be a long-lasting and energy-efficient success.

Chapter 10

Combined Heat and Power With Microturbines

In combined heat and power (CHP) installations, a turbine or turbine system (large industrial or microturbine-sized) produces power via its main shaft, which drives a generator. Conventional (simple cycle) power plants frequently emit the heat created as a by-product of power into the environment as flue gas. CHP captures the excess heat for domestic or industrial heating purposes, either very close to the plant, or—(commonly in eastern Europe)—distributes heat via steam pipes to heat local housing (i.e., "district heating"). CHP is therefore an example of a cogeneration (withwaste heat recovery) cycle.

The exhaust heat from the cycle can serve a variety of purposes. By capturing the excess heat, CHP allows a more total use of energy than conventional generation, potentially reaching an efficiency of 70% to 90%, compared with approximately 50% for the best conventional simple cycle plants.

The use of CHP is limited by the fact that, although it is more efficient (than using heat generated from other additional sources) if the heat can be used on-site or very close to it, it is less efficient when the heat needs to be transported over longer distances. Heat transmission over long distances requires thick, heavily insulated pipes. With well-designed transmission and distribution systems, electricity can be transmitted over longer distances than with heat transmission. The operative phrase here is "well-designed" as electrical transmission systems in countries with poor infrastructure can incur losses as high as 30% of power produced.

The hot exhaust gases may be directly ducted into greenhouses, for instance. This is a common application in cold countries. The hot gases may be ducted through heating system ducting. In many small- to medium-sized towns and villages in Europe, houses and businesses are frequently clustered around their power-generation plants so that the hot air energy losses are minimized.

Cogeneration plants also occur in the district heating systems of big towns, universities, hospitals, hotels, prisons, oil refineries, paper mills, wastewater treatment plants, enhanced oil recovery wells, and industrial plants with large heating needs.

Large or small, most cogeneration projects are designed to produce the energy needs of the facility. However, thermally enhanced oil recovery plants often produce a substantial amount of excess electricity. After generating electricity, these plants inject leftover steam into heavy oil wells so that the oil will flow easier, increasing production. Thermally enhanced oil recovery cogeneration plants in Kern County, CA, produce so much electricity that it cannot all be used locally. The excess is transmitted to Los Angeles.

TYPES OF PLANTS

Topping-cycle plants produce electricity first, then the exhaust is used for heating. Bottoming-cycle plants, which are rare, produce heat for an industrial process first, then electricity is produced using a waste-heat-recovery boiler. Bottoming-cycle plants are only used when the industrial process requires very high temperatures, such as furnaces for glass and metal manufacturing.

Large cogeneration systems provide heating water and power for an industrial site or an entire town. Common CHP plant types include the following:

- Gas turbine CHP plants using the waste heat in the flue gas of gas turbines
- Combined-cycle power plants adapted for CHP
- Steam turbine CHP plants using the waste heat in the steam after the steam turbine

Smaller cogeneration units usually use a reciprocating engine or Stirling engine. They use the waste heat in the flue gas and cooling water of gas or diesel engines and replace the traditional gas- or oil-fired boiler (furnace) used in central heating systems.

MICROCHP

"Micro-cogeneration" is on the scale of one household or small business. Instead of burning fuel to merely heat the house or hot water, some of the energy is converted to electricity in addition to heat. This electricity can be used within the home or business, or—(with a permit from the network operator)—fed back into the network at a profit.

WHAT MAKES A PLANT A GOOD CANDIDATE FOR CHP?*

See Figure 10-1.

MICROTURBINE CHP SYSTEMS*

Microturbines are well suited for distributed generation applications due to their flexibility with connection methods, and ability to be stacked in parallel to serve larger loads. They provide stable and reliable power, and low emissions. Types of applications include:

- Peak shaving and base load power (grid parallel)
- CHP
- Stand-alone power
- Backup/standby power
- Ride-through connection
- Primary power with grid as backup
- Microgrid
- Resource recovery

* Source: Energy Nexus Group. 2002. *Technology Characterization: Microturbines.* U.S. Environmental Protection Agency. Washington, D.C.

STEP 1	Please check the boxes that apply to you:
☐	Do you pay more than $.06/ kWh on average for electricity (including generation, transmission, and distribution)?
☐	Are you concerned about the impact of current or future energy costs on your business?
☐	Is your facility located in a deregulated electricity market?
☐	Are you concerned about power reliability? Is there a substantial financial impact to your business if the power goes out for 1 hour? For 5 minutes?
☐	Does your facility operate for more than 5000 hours/year?
☐	Do you have thermal loads throughout the year (including steam, hot water, chilled water, process heat, etc.)?
☐	Does your facility have an existing central plant?
☐	Do you expect to replace, upgrade, or retrofit central plant equipment within the next 3-5 years?
☐	Do you anticipate a facility expansion or new construction project within the next 3-5 years?
☐	Have you already implemented energy efficiency measures and still have high energy costs?
☐	Are you interested in reducing your facility's impact on the environment?
STEP 2	If you have answered "yes" to 3 or more of these questions, your facility may be a good candidate for CHP.

The next step in assessing the potential of an investment in CHP is to have a Level 1 Feasibility analysis performed to estimate the preliminary return on investment. The EPA CHP Partnership offers comprehensive Level 1 analysis services for qualifying projects and can provide contact information to others who perform these types of analyses.

FIGURE 10-1   *Checklist for potential combined heat and power plants. (Source: http://www. epa.gov/)*

Customers include financial services, data processing, telecommunications, restaurant, lodging, retail, office building, and other commercial sectors. Microturbines currently operate in resource recovery operations at oil and gas production fields, wellheads, coal mines, and landfill operations. In landfills, a system of pipes collect methane produced from the fill, so in this case fuel is "free." Reliable unattended operation is important since these locations may be remote from the grid, and even when served by the grid, may experience costly downtime when electric service is lost due to weather, fire, or animals.

In CHP applications, the waste heat from the microturbine is used to produce hot water, to heat building space, to drive absorption cooling or desiccant dehumidification equipment, and to supply other thermal energy needs in a building or industrial process.

Technology Description

Basic Processes

Microturbines are small gas turbines, most of which feature an internal heat exchanger called a *recuperator*. In a microturbine, a radial flow (centrifugal) compressor compresses the inlet air that is then preheated in the recuperator using heat from the turbine exhaust. Next, the heated air from the recuperator mixes with fuel in the combustor and hot gases (products of combustion) expand through the expansion and power turbines.

The expansion turbine turns the compressor and, in single-shaft models, turns the generator as well. Two-shaft models may use a second turbine module to drive the generator, generally via a gearbox. Finally, the recuperator uses the exhaust of the power turbine to preheat discharge air from the compressor.

Single-shaft models generally operate at speeds over 60,000 revolutions per minute (RPM) and generate electric power of high frequency and of variable frequency (alternating current [AC]). This power is rectified to direct current (DC) and then inverted to 60 hertz (Hz) for U.S. commercial use.

In the two-shaft version, the power turbine connects via a gearbox to a generator that produces power at 60 Hz. Some manufacturers offer units producing 50 Hz for use in countries where 50 Hz is standard, such as in Europe and parts of Asia.

Thermodynamic Cycle.   As seen previously, microturbines are essentially small gas turbines. They operate on the same thermodynamic cycle, known as the Brayton cycle, as larger gas turbines. In this cycle, atmospheric air is compressed, heated, and then expanded. The excess power produced by the turbine expander in excess of the compressor's needs, is used for power generation.

As with conventional gas turbines, the power produced by an expansion turbine and consumed by a compressor is proportional to the absolute temperature of the gas passing through those devices. Consequently, turbine inlet temperatures are designed for the highest practical temperature consistent with economic materials. Compressor inlet airflow temperature is kept as low as possible.

As technology advances permit higher turbine inlet temperature, optimum pressure ratio also increases. Higher temperature and pressure ratios result in higher efficiency and specific power.

The general trend in gas turbine advancement has been toward a combination of higher temperatures and pressures. However, microturbine inlet temperatures are generally limited to 1800°F or below to enable the use of relatively inexpensive materials for the turbine wheel and to maintain pressure ratios at a comparatively low 3.5 to 4.0.

Basic Components

Turbo-Compressor Package.   As seen previously, the basic components of a microturbine are the compressor, turbine generator, and recuperator (Figure 10-2). The heart of the microturbine is the compressor-turbine package, which is commonly mounted on a single shaft along with the electric generator. Two bearings support the single shaft. The single moving part of the one-shaft design has the potential for reducing maintenance needs and enhancing overall reliability.

There are two-shaft versions, in which the turbine on the first shaft directly drives the compressor while a power turbine on the second shaft drives a gearbox and conventional electrical generator producing 60 Hz power. The two-shaft design features more moving parts but does not require complicated power electronics to convert high-frequency AC power output to 60 Hz.

Moderate- to large-size gas turbines use multistage axial flow turbines and compressors, in which the gas flows along the axis of the shaft and is compressed and expanded in multiple stages. However, microturbine turbomachinery is based on single-stage radial flow compressors and turbines.

Radial flow turbomachinery handles the small volumetric flows of air and combustion products with reasonably high component efficiency. (Axial flow machines move greater volumes and blade height would be too low to be practical).

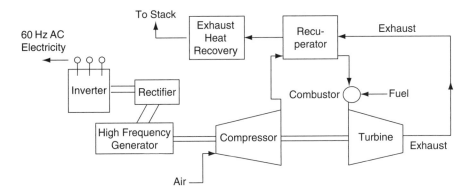

FIGURE 10-2 *Microturbine-based combined heat and power system (single-shaft design).* *(Source: Energy Nexus Group. 2002.* Technology characterization: microturbines. *U.S. Environmental Protection Agency. Washington, D.C.)*

Large-size axial flow turbines and compressors are typically more efficient than radial flow components. However, in the size range of microturbines—0.5 to 5 pounds (lbs) per second of air-gas flow—radial flow components offer minimum surface and end wall losses and provide the highest efficiency.

In microturbines, the turbocompressor shaft generally turns at about 96,000 RPM in the case of a 30-kilowatt (kW) machine and about 80,000 RPM in a 75-kW machine. One 45-kW model on the market turns at 116,000 RPM. There is no single rotational speed-power size rule, as the specific turbine and compressor design characteristics strongly influence the physical size of components and consequently rotational speed. For a specific aerodynamic design, as the power rating decreases, the shaft speed increases, hence the high shaft speed of the small microturbines.

The radial flow turbine-driven compressor is quite similar in terms of design and volumetric flow to automobile, truck, and other small reciprocating engine turbochargers. Superchargers and turbochargers are used to increase the power of reciprocating engines by compressing the inlet air to the engine. Today's world market for small automobile and truck turbochargers is around two million units per year.

Small gas turbines, of the size and power rating of microturbines, serve as auxiliary power units (APUs) on airplanes. Cabin cooling (air conditioning) systems of airplanes use this same size and design family of compressors and turbines. The decades of experience with these applications provide the basis for the engineering and manufacturing technology of microturbine components.

Generator. As seen previously, the microturbine produces electrical power either via a high-speed generator turning on the single turbo-compressor shaft or with a separate power turbine driving a gearbox and conventional 3600-RPM generator. The high-speed generator of the single-shaft design employs a permanent magnet (typically samarium-cobalt) alternator and requires that the high-frequency AC output (about 1600 Hz for a 30-kW machine) be converted to 60 Hz for general use.

This power conditioning involves rectifying the high-frequency AC to DC, and then inverting the DC to 60 Hz AC. Power conversion comes with an efficiency penalty (approximately 5%). To start up a single-shaft design, the generator acts as a motor turning the turbo-compressor shaft until sufficient RPM is reached to start the combustor. If the system is operating independent of the grid (black starting), a power storage unit (typically an uninterrupted-power-supply battery) is used to power the generator for start-up.

Recuperators.   Recuperators are heat exchangers. Hot turbine exhaust gas (typically around 1200°F) preheats the compressed air (typically around 300°F) going into the combustor, reducing the fuel needed to heat the compressed air to turbine inlet temperature. Depending on microturbine operating parameters, recuperators can more than double machine efficiency.

Since there is increased pressure drop in both the compressed air and turbine exhaust sides of the recuperator, power output typically declines 10% to 15% from that attainable without the recuperator. Recuperators also lower the temperature of the microturbine exhaust, reducing the microturbine's effectiveness in CHP applications.

Bearings.   Microturbines operate on either oil-lubricated or air bearings, which support the shaft(s). *Oil-lubricated* bearings are mechanical bearings and come in three main forms—high-speed metal roller, floating sleeve, and ceramic surface. The latter typically has the best life, operating temperature, and lubricant flow.

While oil-lubricated bearings are a well-established technology, they require an oil pump, oil filtering system, and liquid cooling that add to cost and maintenance. Also the exhaust from machines featuring oil-lubricated bearings may not be useable for direct space heating in cogeneration configurations due to the potential for contamination. Since the oil never comes in direct contact with hot combustion products, as is the case in small reciprocating engines, the reliability of such a lubrication system is more typical of ship propulsion diesel systems (which have separate bearings and cylinder lubrication systems) and automotive transmissions than cylinder lubrication in automotive engines.

Air bearings have been in service on airplane cabin cooling systems for many years. They allow the turbine to spin on a thin layer of air, so friction is low and RPM is high. No oil or oil pump is needed.

Air bearings are simple to operate, without the cost, reliability concerns, maintenance requirements, or power drain of an oil supply and filtering system. Concern does exist for the reliability of air bearings under numerous and repeated starts due to metal on metal friction during start-up, shutdown, and load changes. Reliability depends more on an OEM's quality control methodology than on design, which is proven only after significant experience with many units that have extended operating hours and start cycles.

Power Electronics.   As discussed previously, single-shaft microturbines feature digital power controllers to convert the high-frequency AC power produced by the generator into usable electricity. The high-frequency AC is rectified to DC, inverted back to 60 or 50 Hz AC, and then filtered to reduce harmonic distortion. This is a critical component in the single-shaft microturbine design and represents significant design challenges, specifically in matching turbine output to the required load. To allow for transients and voltage spikes, power electronics designs are generally able to handle seven times the nominal voltage. Most microturbine power electronics generate three-phase electricity.

Electronic components also control operating and start-up functions. Microturbines are generally equipped with controls that allow the unit to be operated in parallel or independent of the grid. They internally incorporate many of the grid and system protection features required for interconnect. The controls also allow for remote monitoring and operation.

CHP Operation

In CHP operation, a second heat exchanger, the exhaust gas heat exchanger, transfers the remaining energy from the microturbine exhaust to a hot water system. Exhaust heat can be used for a number of different applications, including potable water heating, driving absorption cooling and desiccant dehumidification equipment, space heating, process heating, and other building or site uses.

Some microturbine-based CHP applications do not use recuperators. With these microturbines, the temperature of the exhaust is higher and thus more heat is available for recovery. Figure 10-2 illustrates a microturbine-based CHP system.

Design Characteristics

- Thermal output: Microturbines produce exhaust gases at temperatures in the 400°F to 600°F range, suitable for supplying a variety of building thermal needs.
- Fuel flexibility: Microturbines can operate using a number of different fuels: natural gas, sour gases (high sulfur, low BTU content), and liquid fuels such as gasoline, kerosene, and diesel fuel/heating oil.
- Reliability and life: Design life is estimated to be in the 40,000- to 80,000-hour range. While units have demonstrated reliability, the global fleet has not been in commercial service long enough to provide definitive life data.
- Size range: Microturbines available and under development are sized from 30 to 350kW.
- Emissions: Low inlet temperatures and high fuel-to-air ratios result in oxides of nitrogen (NO_x) emissions of less than 10 parts per million (ppm) when running on natural gas.
- Modularity: Units may be connected in parallel to serve larger loads and provide power reliability.
- Part-load operation: Because microturbines reduce power output by reducing mass flow and combustion temperature, efficiency at part load can be below that of full-power efficiency.
- Dimensions: Dimensions are about 12 cubic feet.

Performance Characteristics

As noted earlier, the microturbine cycle is more complex than that of a conventional simple-cycle gas turbine, as the addition of the recuperator both:

- Reduces fuel consumption (thereby substantially increasing efficiency) and
- Introduces additional internal pressure losses that moderately lower efficiency and power.

As the recuperator has four connections—to the compressor discharge, the expansion turbine discharge, the combustor inlet, and the system exhaust—it becomes a challenge to the microturbine product designer to make all of the connections in a manner that minimizes pressure loss, keeps manufacturing cost low, and entails the least compromise of system reliability. Each manufacturer's models have evolved in unique ways.

The addition of a recuperator opens numerous design parameters to performance-cost trade-offs. In addition to selecting the pressure ratio for high efficiency and best business opportunity (high power for low price), the recuperator has two performance parameters, effectiveness and pressure drop, that also have to be selected for optimum efficiency and cost.

Higher effectiveness recuperation requires greater recuperator surface area, which both increases cost and incurs additional pressure drop. Such increased internal pressure drop reduces net power production and increases microturbine cost per kW.

Microturbine performance, in terms of both efficiency and specific power (power produced per unit of mass flow) is highly sensitive to small variations in component performance and internal losses. This is because the high-efficiency recuperated cycle processes a much larger amount of air and combustion products flow per kW of net powered delivered than is the case for high-pressure ratio simple-cycle machines. When the net output is the small difference between two large numbers (the compressor and expansion turbine work per unit of mass flow), small losses in component efficiency, internal pressure losses, and recuperator effectiveness have large impacts on net efficiency and net power per unit of mass flow.

Electrical Efficiency

Figure 10-3 shows a recuperated microturbine electrical efficiency as a function of microturbine compressor ratio, for a range of turbine firing temperatures from 1550°F to 1750°F, corresponding to conservative to optimistic turbine material life behavior. The nominal efficiency is the gross generator output (without parasitic or conversion losses considered). Often this is at high frequency, so the output must be rectified and inverted to provide 60 Hz AC power.

The efficiency loss in such frequency conversion (about 5%, which would lower efficiency from 30% to 28.5%) is not included in these charts. Figure 10-3 shows that a broad optimum of performance exists in the pressure ratio range from 3 to 4.

Figure 10-4 shows microturbine-specific power for the same range of firing temperatures and pressure ratios. Higher pressure ratios result in greater specific power. However, practical considerations limit compressor and turbine component highest rotational speeds due to centrifugal forces and allowable stresses in economic materials, resulting in compressor pressure ratio limits of 3.5 to 5 in microturbines currently entering the market.

Table 10-1 summarizes performance characteristics for typical microturbine CHP systems. The range of 30 to 350 kW represents what is currently or soon to be commercially available. Heat rates and efficiencies shown were taken from

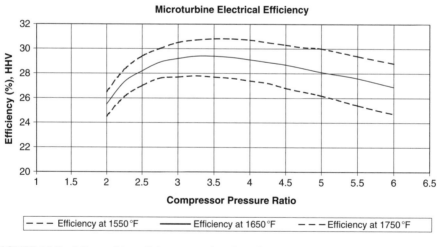

FIGURE 10-3 *Microturbine efficiency as a function of compressor pressure ratio and turbine firing temperature. Most of the efficiencies quoted in this report are based on higher heating value (HHV), which includes the heat of condensation of the water vapor in the combustion products. In engineering and scientific literature, the lower heating value (LHV) is often used, which does not include the heat of condensation of the water vapor in the combustion products. Fuel is sold on an HHV basis. The HHV is greater than the LHV by approximately 10% with natural gas as the fuel (i.e., 50% LHV is equivalent to 45% HHV). HHV efficiencies are about 8% greater for oil (liquid petroleum products) and 5% greater for coal. (Source: Energy Nexus Group. 2002.* Technology characterization: microturbines. *U.S. Environmental Protection Agency. Washington, D.C.)*

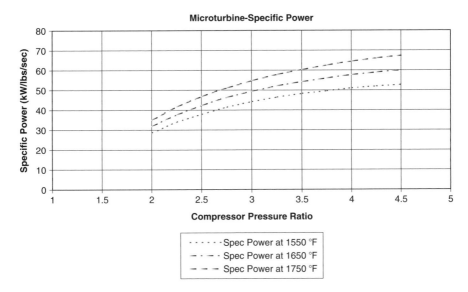

FIGURE 10-4 *Microturbine-specific power as a function of compressor pressure ratio and turbine firing temperature. (Source: Energy Nexus Group. 2002.* Technology characterization: microturbines. *U.S. Environmental Protection Agency. Washington, D.C.)*

manufacturers' specifications and industry publications. Electrical efficiencies are the net of parasitic and conversion losses.

Available thermal energy is calculated based on manufacturer specifications on turbine exhaust flows and temperatures. CHP thermal recovery estimates are based on producing hot water for process or space-heating applications. Total CHP efficiency is the sum of the net electricity generated plus hot water produced for building thermal needs divided by total fuel input to the system.

Effective electrical efficiency is a more useful value than overall efficiency to measure fuel savings. Effective electric efficiency assumes that a water heater would otherwise generate the useful thermal output from the CHP system at an 80% thermal efficiency. The theoretical water heating fuel use is subtracted from the total fuel input to calculate the effective electric efficiency of the CHP system.

Data in the table show that electrical efficiency increases as the microturbine becomes larger. As electrical efficiency increases, the absolute quantity of thermal energy available decreases per unit of power output, and the ratio of power to heat for the CHP system increases. A changing ratio of power to heat impacts project economics and may affect the decisions that customers make in terms of CHP acceptance, sizing, and other characteristics.

Each microturbine manufacturer represented in Table 10-1 uses a different recuperator, and each has made individual trade-offs between cost and performance. Performance involves the extent to which the recuperator effectiveness increases cycle efficiency, the extent to which the recuperator pressure drop decreases cycle power, and the choice of what cycle pressure ratio to use. So microturbines of different makes will have different CHP efficiencies and different net heat rates chargeable to power.

As shown, microturbines typically require 50 to 80 psig fuel supply pressure. Because microturbines are built with pressure ratios between 3 and 4 to maximize efficiency with a recuperator at modest turbine inlet temperature, the required supply

TABLE 10-1   Microturbine Combined Heat and Power (CHP)—Typical Performance
Parameters[*]

Cost and Performance Characteristics[1]	System 1	System 2	System 3	System 4
Nominal electricity capacity (kW)	30 kW	70 kW	100 kW	350 kW
Package cost ($2000/kW)[2]	$1000	$950	$800	$750
Total installed cost ($2000/kW)[3]	$2516	$2031	$1561	$1339
Electric heat rate (BTU/kWh), HHV[4]	14,581	13,540	12,637	11,766
Electrical efficiency (%), HHV[5]	23.4%	25.2%	27.0%	29.0%
Fuel input (MMBTU/hr)	0.437	0.948	1.264	4.118
Required fuel gas pressure (psig)	55	55	75	135
CHP Characteristics				
Exhaust flow (lbs/sec)	0.72	1.40	1.74	5.00
GT exhaust temp (°F)	500	435	500	600
Heat exchanger exhaust temp (°F)	150	130	131	140
Heat output (MMBTU/hr)	0.218	0.369	0.555	1.987
Heat output (kW equivalent)	64	108	163	582
Total CHP efficiency (%), HHV[6]	73%	64%	71%	77%
Power/heat ratio[7]	0.47	0.65	0.62	0.60
Net heat rate (BTU/kWh)[8]	5509	6952	5703	4668
Effective electrical efficiency (%), HHV[9]	62%	49%	60%	73%

GT = gas turbine.

[*] For typical systems commercially available in 2001 (30-, 70-, and 100-kW units) or soon to be available
(350-kW model is under development). 30-, 100-, and 350-kW systems represented are single-shaft models.
70-kW system represented is a double-shaft model.

[1] Characteristics presented are representative of "typical" commercially available or soon to
be available microturbine systems. Table data are based on: Capstone Model 330—30 kW; IR
Energy Systems 70LM—70 kW (two-shaft); Turbec T100—100 kW; DTE model currently under
development—350 kW.

[2] Equipment cost only. The cost for all units except for the 30-kW unit includes integral heat recovery water
heater. All units include a fuel gas booster compressor.

[3] Installed costs based on CHP system producing hot water from exhaust heat recovery. The 70-kW and
100-kW systems are offered with integral hot water recovery built into the equipment. The 30-kW units are
currently built as electric (only) generators and the heat recovery water heater is a separate unit. Other units
entering the market are expected to feature built-in heat recovery water heaters.

[4] All turbine and engine manufacturers quote heat rates in terms of the lower heating value (LHV) of the
fuel. On the other hand, the usable energy content of fuels is typically measured on a higher-heating-value
(HHV) basis. In addition, electric utilities measure power plant heat rates in terms of HHV. For natural gas,
the average heat content of natural gas is 1030 BTU/scf on an HHV basis and 930 BTU/scf on an LHV
basis—or about a 10% difference.

[5] Electrical efficiencies are the net of parasitic and conversion losses. Fuel gas compressor needs based on
1 psi inlet supply.

[6] Total efficiency = (net electric generated + net heat produced for thermal needs)/total system fuel input

[7] Power/heat ratio = CHP electrical power output (BTU)/ useful heat output (BTU)

[8] Net heat rate = (total fuel input to the CHP system – the fuel that would be normally used to generate the
same amount of thermal output as the CHP system output assuming an efficiency of 80%) /CHP electric
output (kW).

[9] Effective Electrical Efficiency = (CHP electric power output)/(Total fuel into CHP system – total heat
recovered/0.8).

(Source: Energy Nexus Group. 2002. *Technology characterization: microturbines*. U.S. Environmental
Protection Agency. Washington, D.C.)

pressure for microturbines is much less than for industrial-size gas turbines with pressure ratios of 7 to 35.

Local distribution gas pressures usually range from 30 to 130 psig in feeder lines and from 1 to 50 psig in final distribution lines. Most U.S. businesses that would use a 30-, 70-, or 100-kW microturbine receive gas at about 0.5 to 1.0 psig. Also most building codes prohibit piping higher-pressure natural gas within the structure. Thus, microturbines in most commercial locations require a fuel gas booster compressor to ensure that fuel pressure is adequate for the gas turbine flow control and combustion systems.

Most microturbine manufacturers offer the equipment package with the fuel gas booster included. It is included in all of the representative systems shown in Table 10-1. This packaging facilitates the purchase and installation of a microturbine, as the burden of obtaining and installing the booster compressor is no longer placed on the customer. Also, it might result in higher reliability of the booster through standardized design and volume manufacture.

Booster compressors can add from $50 to $100 per kW to a microturbine CHP system's total cost. As well as adding to capital cost, booster compressors lower net power and efficiency so the operating cost is slightly higher. Typically, the fuel gas booster requires about 5% of the microturbine output. For example, a single 60-kW unit requires 2.6 kW for the booster, while a booster serving a system of three 30-kW units would require 4.4 kW. Such power loss results in an efficiency penalty of about 1.5%. For installations where the unit is located outdoors, the customer can save on cost and operating expense by having the gas utility deliver gas at an adequate pressure and obtaining a system without a fuel gas booster compressor.

Part-Load Performance

When less than full power is required from a microturbine, the output is reduced by a combination of mass flow reduction (achieved by decreasing the compressor speed) and turbine inlet temperature reduction. In addition to reducing power, this change in operating conditions also reduces efficiency. Figure 10-5 shows a sample part-load derate curve for a microturbine.

Effects of Ambient Conditions on Performance

The ambient conditions under which a microturbine operates have a noticeable effect on both the power output and efficiency. At elevated inlet air temperatures, both the power and efficiency decrease. The power decreases due to the decreased airflow mass rate (since the density of air declines as temperature increases), and the efficiency decreases because the compressor requires more power to compress air of higher temperature.

Conversely, the power and efficiency increase with reduced inlet air temperature. Figure 10-6 shows the variation in power and efficiency for a microturbine as a function of ambient temperature compared to the reference International Organization for Standards condition of sea level and 59°F. The density of air decreases at altitudes above sea level. Consequently, power output decreases. Figure 10-7 illustrates the altitude derate.

Heat Recovery

Effective use of the thermal energy contained in the exhaust gas improves microturbine system economics. Exhaust heat can be recovered and used in a variety of ways, including water heating, space heating, and driving thermally activated equipment such as an absorption chiller or a desiccant dehumidifier.

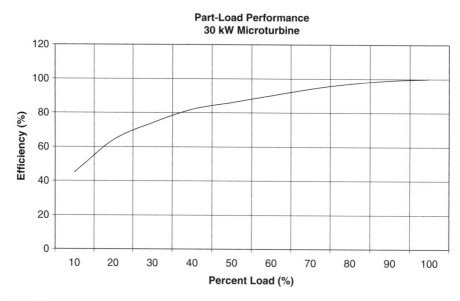

FIGURE 10-5   *Microturbine part-load power performance.* Note: *Unit represented is a single-shaft, high-speed alternator system. (Source: Energy Nexus Group. 2002.* Technology characterization: microturbines. *U.S. Environmental Protection Agency. Washington, D.C.)*

Microturbine CHP system efficiency is a function of exhaust heat temperature. Recuperator effectiveness strongly influences the microturbine exhaust temperature. Consequently, the various microturbine CHP systems have substantially different CHP efficiency and net heat rate chargeable to power. These variations in CHP efficiency and net heat rate are mostly due to the mechanical design and manufacturing cost of the recuperators and their resulting impact on system cost, rather than being due to differences in system size.

Performance and Efficiency Enhancements

Recuperators.   Most microturbines include built-in recuperators. The inclusion of a high-effectiveness (90%)[*] recuperator essentially doubles the efficiency of a microturbine with a pressure ratio of 3.2, from about 14% to about 29% depending on component details. Without a recuperator, such a machine would be suitable only for emergency, backup, or possibly peaking power operation. With the addition of the recuperator, a microturbine can be suitable for intermediate duty or price-sensitive baseload service.

While recuperators previously in use on industrial gas turbines developed leaks attributable to the consequences of differential thermal expansion accompanying thermal transients, microturbine recuperators have proven quite durable in testing to date. This durability has resulted from using higher-strength alloys and higher-quality welding along with engineering design to avoid the internal differential expansion that causes internal stresses and leakage.

Such practical improvements result in recuperators being of appreciable cost, which detracts from the economic attractiveness of the microturbine. The cost of

---

[*] *Effectiveness* is the technical term in the heat-exchanger industry for the ratio of the actual heat transferred to the maximum achievable.

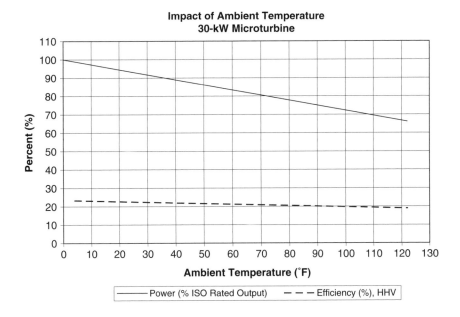

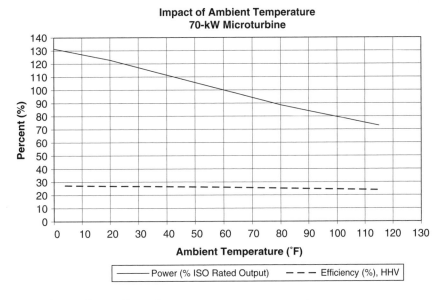

Source: Energy Nexus Group.

FIGURE 10-6 *Ambient temperature effects on microturbine performance. ISO = International Organization for Standards; HHV = higher heating value. (Source: Energy Nexus Group. 2002. Technology characterization: microturbines. U.S. Environmental Protection Agency. Washington, D.C.)*

a recuperator becomes easier to justify as the number of full-power operational hours per year increases.

Incorporation of a recuperator into the microturbine results in pressure losses in the recuperator itself and in the ducting that connects it to other components. Typically, these pressure losses result in 10% to 15% less power being produced by the microturbine

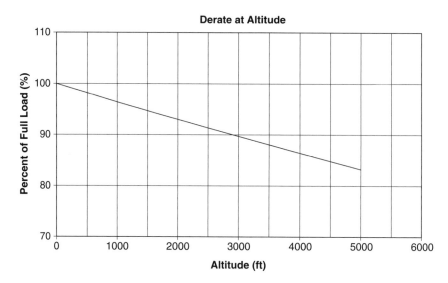

FIGURE 10-7   *Altitude effects on microturbine performance. (Source: Energy Nexus Group. 2002. Technology characterization: microturbines. U.S. Environmental Protection Agency. Washington, D.C.)*

and a corresponding loss of a few points in efficiency. The pressure loss parameter in gas turbines that is the measure of lost power is δp/p. As δp/p increases, the net pressure ratio available for power generation decreases, and hence the power capability of the expansion process diminishes as well. Figure 10-8 illustrates the relationship between recuperator effectiveness and microturbine efficiency.

Firing Temperature.   Large turbines (25 to 2000 lbs per second of mass flow) are usually equipped with internal cooling capability for flame temperatures well above those of the metallurgical limit of the best gas turbine alloys. So progress to higher gas turbine efficiency, via higher firing temperatures, has occurred more through the development and advancement of blade and vane internal cooling technology than through the improvement of the high-temperature capabilities of gas turbine alloys.

Unfortunately for microturbine development, the nature of the three-dimensional shape of radial inflow turbines has not yet lent itself to the development of a manufacturing method that can produce the internal cooling passages common in high performance gas turbine axial turbine blades. Consequently, microturbines are limited to firing temperatures within the capabilities of gas turbine alloys.

An ongoing program at the U.S. Department of Energy (DOE) Office of Energy Efficiency seeks to apply the technology of ceramic radial inflow turbines (previously advanced for the purpose of developing automotive gas turbines) to microturbines, to increase their efficiency to 36% (HHV). The design and materials technology from the previous efforts are applicable, since the automotive gas turbines were in the same size range, and of the same general geometry, as those used in microturbines.

Inlet Air Cooling.   As shown in Figure 10-6, the decreased power and efficiency of microturbines at high ambient temperatures means that microturbine performance is at its lowest at the times power is often in greatest demand and most valued. The use of inlet air cooling can mitigate the decreased power and efficiency resulting from high ambient air temperatures.

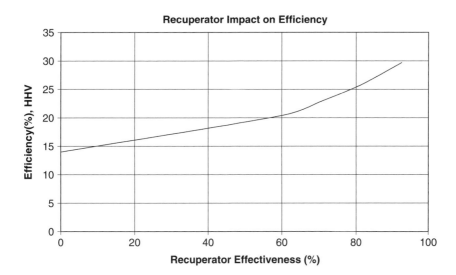

FIGURE 10-8 *Microturbine efficiency as a function of recuperator effectiveness. (Source: Energy Nexus Group. 2002.* Technology characterization: microturbines. *U.S. Environmental Protection Agency. Washington, D.C.)*

While inlet air cooling is not a feature on today's microturbines, cooling techniques now entering the market on large gas turbines can be expected to work their way to progressively smaller equipment sizes. At some future date, they will be used with microturbines.

Evaporative cooling, a relatively low-capital-cost technique, is the most likely to be applied to microturbines. It uses a very fine spray of water directly into the inlet air stream. Evaporation of the water reduces the temperature of the air.

Since cooling is limited to the wet-bulb air temperature, evaporative cooling is most effective when the wet-bulb temperature is appreciably below the dry-bulb (i.e., ordinary) temperature. In most locales with high daytime dry-bulb temperatures, the wet-bulb temperature is often 20°F lower. This affords an opportunity for substantial evaporative cooling. However, evaporative cooling can consume large quantities of water, making it difficult to operate in arid climates.

Refrigeration cooling in microturbines is also technically feasible. In refrigeration cooling, a compression-driven or thermally activated (absorption) refrigeration cycle cools the inlet air through a heat exchanger. The heat exchanger in the inlet air stream causes an additional pressure drop in the air entering the compressor, thereby slightly lowering cycle power and efficiency.

However, as the inlet air is now cooler than the ambient air, there is a significant net gain in power and efficiency. Electric motor compression refrigeration requires a substantial parasitic power loss. Thermally activated absorption cooling can use waste heat from the microturbine, reducing the direct parasitic loss. The complexity and cost of these approaches, in comparison with evaporative cooling, leaves them at a disadvantage.

Finally, one can use thermal energy storage systems, typically ice, chilled water, or low-temperature fluids, to cool inlet air. These systems eliminate most parasitic losses from the augmented power capacity.

Thermal energy storage is a viable option if on-peak power pricing only occurs a few hours a day. In that case, the shorter time of energy storage discharge and longer time for daily charging allow for a smaller and less expensive thermal energy storage system.

Capital Cost

This section provides typical study estimates for the installed cost of microturbine systems. Two configurations are presented: power-only and CHP producing hot water for use on-site. Equipment-only and installed costs are estimated for the four typical microturbine systems.

These are "typical" budgetary price levels. It should also be noted that installed costs can vary significantly depending on the scope of the plant equipment, geographical area, competitive market conditions, special site requirements, emissions control requirements, prevailing labor rates, and whether the system is a new or retrofit application.

Table 10-2 provides cost estimates for combined heat and power applications, assuming that the CHP system produces hot water. The basic microturbine package consists of the turbogenerator package and power electronics. All of the commercial and near-commercial units offer basic interconnection and paralleling functionality as part of the package cost. All but one of the systems offers an integrated heat exchanger heat recovery system for CHP within the package.

All OEMs also indicate that the package price includes the gas booster compressor. It should be noted that the package prices cited in Table 10-2 represent manufacturer quotes or estimates. However, only two of the products have any market history, while the other two are planning to enter the market this year. The manufacturer quotes may not reflect actual cost plus profit today, but may instead represent a *forward pricing* strategy in which early units are sold at a loss to develop the market. The information provided for each sample system is as follows:

- 30 kW: Single unit $1000/kW, including fuel gas compressor, DC-to-AC inverter, all electronic interconnection hardware, but without the heat recovery heat exchanger. Prices are lower for volume purchases, which are favored. (Capstone: nearly 2000 units shipped to dealer network.)

TABLE 10-2   Estimated Capital Cost for Microturbine Generators in Grid-Interconnected Combined Heat and Power Application

Cost Component	System 1	System 2	System 3	System 4
Nominal capacity (kW)	30	70	100	350
Costs ($/kW)				
Equipment				
Microturbine	$1000	$1030	$800	$750
Gas booster compressor	incl.	incl.	incl.	incl.
Heat recovery	$225	incl.	incl.	incl.
Controls/monitoring	$179	$143	$120	$57
Total equipment	$1403	$1173	$920	$807
Labor/materials	$429	$286	$200	$160
Total process capital	$1832	$1459	$1120	$967
Project and construction management	$418	$336	$260	$226
Engineering and fees	$154	$146	$112	$86
Project contingency	$72	$58	$45	$38
Project financing (interest during construction)	$40	$32	$25	$21
Total plant cost ($/kW)	$2516	$2031	$1561	$1339

incl. = cost included.
(Source: Energy Nexus Group. 2002. *Technology characterization: microturbines*. U.S. Environmental Protection Agency. Washington, D.C.)

- 70 kW: Price of $79,900 includes commissioning and the first year of maintenance (at $0.01/kWh and 4000 hours/year, equivalent to $40/kW). Built-in heat recovery heat exchanger included in price. Generator is a standard 3600-RPM AC unit; therefore, there is no need for an inverter. Electrical interconnection and fuel gas booster compressor are included. For this comparison, prepaid maintenance and commissioning costs have been backed out from the package price.
- 100 kW: A price of $800/kW is offered to distributors for equipment including heat recovery heat exchanger (built-in), fuel gas booster, DC-to-AC inverter, and all interconnection hardware.
- 350 kW: Price target of $910/kW for all equipment, including heat recovery heat exchanger, inverter, fuel gas booster and interconnection, installed. For this comparison, the total package was separated into a package price plus labor.

There is little additional equipment that is required for these integrated systems. A heat recovery system has been added where needed, and additional controls and remote monitoring equipment have been added. The total plant cost consists of total equipment cost plus installation labor and materials (including site work), engineering, project management (including licensing, insurance, commissioning, and start-up), and financial carrying costs during the 6- to 18-month construction period.

The basic equipment costs represent material on the loading dock, ready to ship. The cost to a customer for installing a microturbine-based CHP system includes a number of other factors that increase the total costs by 70% to 80%. Labor/materials represent the labor cost for the civil, mechanical, and electrical work and materials such as ductwork, piping, and wiring. *Total process capital* is the equipment costs plus installation labor and materials.

A number of other costs are incurred on top of total process capital. These costs are often referred to as *soft costs* because they vary widely by installation, by development channel, and by approach to project management. Engineering costs are required to design the system and integrate it functionally with the application's electrical and mechanical systems.

In this characterization, environmental permitting fees are included here. Project and construction management also includes general contractor markup and bonding and performance guarantees. Contingency is assumed to be 3% of the total equipment cost in all cases. Up-front, financing costs are also included.

Since heat recovery is not required for systems that are power-only, the capital costs are lower. For the units that integrate this equipment into the basic package, the savings will be modest, about a $50/kW reduction in the basic package price. Also, installation labor and materials costs are reduced because there is no need to connect with the application's thermal system or to connect the heat recovery equipment in the case where it is a separate unit.

Power-only systems require less engineering time, as integration is required only with the application's electrical system. Project management and construction fees also tend to be lower because it is a more competitive business than CHP. Table 10-3 shows the power-only cost estimates.

As an emerging product, the capital costs shown in Tables 10-2 and 10-3 represent the cost for the early market entry product, though not the cost of the first units into the market. All of the microturbine developer/manufacturers have cost-reduction plans and performance-enhancing developments for the mature market product.

Maintenance

Microturbines are still on a learning curve in terms of maintenance, as initial commercial units have seen only 2 to 3 years of service so far. With relatively few operating hours

TABLE 10-3   Estimated Capital Cost for Microturbine Generators in Grid-Interconnected Power-Only Application

Cost Component	System 1	System 2	System 3	System 4
Nominal capacity (kW)	30	70	100	350
Costs ($/kW)				
Equipment				
Microturbine	$1000	$980	$750	$700
Gas booster compressor	$0	$0	$0	$0
Heat recovery	$0	$0	$0	$0
Controls/monitoring	$179	$143	$120	$57
Total equipment	$1179	$1123	$870	$757
Labor/materials	$300	$200	$140	$112
Total process capital	$1479	$1323	$1010	$869
Project and construction management	$266	$245	$188	$206
Engineering and fees	$130	$85	$64	$44
Project contingency	$56	$50	$38	$34
Project financing (interest during construction)	$31	$27	$21	$18
Total plant cost ($/kW)	$1962	$1729	$1320	$1171

(Source: Energy Nexus Group. 2002. *Technology characterization: microturbines*. U.S. Environmental Protection Agency. Washington, D.C.)

logged as a group, the units in the field have not yet yielded enough data to allow much definition in the area of maintenance.

Most manufacturers offer service contracts for maintenance priced at about $0.01/kWh. This includes periodic inspections of the combustor (and associated hot section parts) and the oil bearing in addition to regular air and oil filter replacements. There have been microturbines operating in environments with extremely dusty air that have required frequent air filter changes due to the dust in the air.

A gas microturbine overhaul is needed every 20,000 to 40,000 hours depending on manufacturer, design, and service. A typical overhaul consists of replacing the main shaft with the compressor and turbine attached and inspecting and, if necessary, replacing the combustor.

At the time of the overhaul, other components are examined to determine whether wear has occurred, with replacements made as required. Microturbines are usually operated with at least one on-off cycle per day. There is concern about the effects of this type of operation on component durability.

There is no known difference in maintenance for operation on fuels other than natural gas. However, experience with liquid fuels in industrial gas turbines suggests that liquid-fueled combustors require more frequent inspections and maintenance than natural gas–fueled combustors.

Fuels

Microturbines have been designed to use natural gas as their primary fuel. However, they are able to operate on a variety of fuels, including the following:

- Liquefied petroleum gas: Propane and butane mixtures
- Sour gas: Unprocessed natural gas as it comes directly from the gas well
- Biogas: Any of the combustible gases produced from biological degradation of organic wastes, such as landfill gas, sewage digester gas, and animal waste digester gas
- Industrial waste gases: Flare gases and process off-gases from refineries, chemical plants, and steel mills
- Manufactured gases: Typically low- and medium-BTU gas produced as products of gasification or pyrolysis processes

Contaminants are a concern with some waste fuels, specifically acid gas components (hydrogen sulfide [H_2S]; halogen acids; hydrogen cyanide [HCN]; ammonia; salts and metal-containing compounds; organic halogen-, sulfur-, nitrogen-, and silicon-containing compounds) and oils. In combustion, halogen and sulfur compounds form halogen acids, sulfur dioxide (SO_2), some sulfites (SO_3), and possibly sulfuric acid (H_2SO_4) emissions. The acids can also corrode downstream equipment.

A fraction of any fuel produces NO_x in combustion. Solid particulates must be kept to low concentrations to prevent corrosion and erosion of components. Various fuel scrubbing, droplet separation, and filtration steps will be required if any fuel contaminant levels exceed manufacturer specifications. Landfill gas in particular often contains chlorine compounds, sulfur compounds, organic acids, and silicon compounds, which dictate pretreatment.

Availability

With the small number of units in commercial service, information is not yet sufficient to draw conclusions about reliability and availability of microturbines. The basic design and low number of moving parts hold the potential for systems of high availability; manufacturers have targeted availabilities of 98% to 99%. The use of multiple units or backup units at a site can further increase the availability of the overall facility.

Emissions

Microturbines have the potential for extremely low emissions. All microturbines operating on gaseous fuels feature lean premixed (dry low NO_x [DLN]) combustor technology, which was developed relatively recently in the history of gas turbines and is not universally featured on larger gas turbines.

The primary pollutants from microturbines are NO_x, carbon monoxide (CO), and unburned hydrocarbons. They also produce a negligible amount of SO_2. Microturbines are designed to achieve the objective of low emissions at full load; emissions are often higher when operating at part load.

The pollutant referred to as NO_x is a mixture of mostly NO and NO_2 in variable composition. In emissions measurement, it is reported as parts per million by volume in which both species count equally. NO_x forms by three mechanisms: thermal NO_x, prompt NO_x, and fuel-bound NO_x. The predominant NO_x formation mechanism associated with gas turbines is thermal NO_x. (Thermal NO_x is the fixation of atmospheric oxygen and nitrogen, which occurs at high combustion temperatures).

Flame temperature and residence time are the primary variables that affect thermal NO_x levels. The rate of thermal NO_x formation increases rapidly with flame temperature. Prompt NO_x forms from early reactions of nitrogen modules in the combustion air and hydrocarbon radicals from the fuel. It forms within the flame, is typically about 1 ppm at 15% oxygen (O_2), and is usually much smaller than the thermal NO_x formation.

Fuel-bound NO_x forms when the fuel contains nitrogen as part of the hydrocarbon structure. Natural gas has negligible chemically bound fuel nitrogen.

Incomplete combustion results in both CO and unburned hydrocarbons. CO emissions result when there is insufficient residence time at high temperature. In gas turbines, the failure to achieve CO burnout may result from combustor wall cooling air. CO emissions are also heavily dependent on operating load. For example, a unit operating under low loads will tend to have incomplete combustion, which will increase the formation of CO. CO is usually regulated to levels below 50 ppm for both health and safety reasons. Achieving such low levels of CO had not been a problem until manufacturers achieved low levels of NO_x, because the techniques used to engineer DLN combustors had a secondary effect of increasing CO emissions.

While not considered a regulated pollutant in the ordinary sense of directly affecting public health, emissions of carbon dioxide (CO_2) are of concern due to its contribution to global warming. Atmospheric warming occurs because solar radiation readily penetrates to the surface of the planet but infrared (thermal) radiation from the surface is absorbed by the CO_2 (and other polyatomic gases such as methane, unburned hydrocarbons, refrigerants, water vapor, and volatile chemicals) in the atmosphere, with resultant increase in temperature of the atmosphere.

The amount of CO_2 emitted is a function of both fuel carbon content and system efficiency. The fuel carbon content of natural gas is 34 lbs carbon/MMBTU; oil is 48 lbs carbon/MMBTU; and (ash-free) coal is 66 lbs carbon/MMBTU.

Lean Premixed Combustion

Thermal NO_x formation is a function of both the local temperatures within the flame and residence time. In older-technology combustors used in industrial gas turbines, fuel and air were separately injected into the flame zone. Such separate injection resulted in high local temperatures at fuel and air zone boundaries.

The focus of combustion improvements of the past decade was to lower flame local hot spot temperature using lean fuel/air mixtures whereby zones of high local temperatures were not created. Lean combustion decreases the fuel/air ratio in the zones where NO_x production occurs so that peak flame temperature is less than the stoichiometric adiabatic flame temperature, therefore suppressing thermal NO_x formation.

All microturbines feature lean premixed combustion systems, also referred to as DLN or dry low emissions. Lean premixed combustion mixes the gaseous fuel and compressed air so that there are no local zones of high temperatures, or "hot spots," where high levels of NO_x would form. DLN requires specially designed mixing chambers and mixture inlet zones to avoid flashback of the flame.

Optimized application of DLN combustion requires an integrated approach to combustor and turbine design. The DLN combustor is an intrinsic part of the turbine design, and specific combustor designs are developed for each turbine application. Full-power NO_x emissions below 9 parts per million by volume (ppmv) at 15% O_2 have been achieved with lean premixed combustion in microturbines.

Catalytic Combustion

In catalytic combustion, fuels oxidize at lean conditions in the presence of a catalyst. Catalytic combustion is a flameless process, allowing fuel oxidation to occur at temperatures below 1700°F, where NO_x formation is low. The catalyst is applied to combustor surfaces, which cause the fuel/air mixture to react on the catalyst surface and release its initial thermal energy. The combustion reaction in the remaining volume of the lean premixed gas then goes to completion at design temperature. Data from ongoing long-term testing indicates that catalytic combustion exhibits low vibration and acoustic noise, only one tenth to one hundredth the levels measured in the same turbine equipped with DLN combustors.

Combustion system and gas turbine developers, along with the U.S. DOE, the California Energy Commission, and other government agencies, are pursuing gas turbine catalytic combustion technology. Past efforts at developing catalytic combustors for gas turbines achieved low, single-digit NO_x ppm levels but failed to produce combustion systems with suitable operating durability. This was typically due to cycling damage and to the brittle nature of the materials used for catalysts and catalyst-support systems. Catalytic combustor developers and gas turbine manufacturers are testing durable catalytic and "partial catalytic" systems that are overcoming the problems of past designs. Catalytic combustors capable of achieving NO_x levels below 3 ppm are in full-scale demonstration and are entering early commercial introduction.[*] As with DLN combustion, optimized catalytic combustion requires an integrated approach to combustor and turbine design. Catalytic combustors must be tailored to the specific operating characteristics and physical layout of each turbine design.

Catalytic combustion may be applied to microturbines as well as industrial and utility turbines. Because of the low emissions from DLN combustors, combined with the low turbine inlet temperatures at which microturbines currently operate, it is not expected that catalytic combustion for microturbines will be pursued in the near term.

TABLE 10-4   Microturbine Emissions Characteristics

Emissions Characteristics[*]	System 1	System 2	System 3	System 4
Nominal electricity capacity (kW)	30	70	100	350
Electrical efficiency, HHV	23%	25%	27%	29%
NO_x, ppmv	9	9	15	9
NO_x, lbs/MWh[*]	0.54	0.50	0.80	0.53
CO, ppmv	40	9	15	25
CO, lbs/MWh	1.46	0.30	0.49	0.72
THC, ppmv	<9	<9	<10	<10
THC, lbs/MWh	<0.19	<0.17	<0.19	<0.19
CO_2 (lbs/MWh)	1928	1774	1706	1529
Carbon (lbs/MWh)	526	484	465	417

THC = tetrahydrachloride.
Note: Estimates are based on manufacturers' guarantees for typical systems commercially available in 2001 (30-, 70-, and 100-kW models). The emissions figures for the 350-kW system under development are manufacturer goals.
[*] Conversion from volumetric emission rate (ppmv at 15% O_2) to output based rate (lbs/MWh) for both NO_x and CO based on conversion multipliers provided by Capstone Turbine Corporation and corrected for differences in efficiency.
(Source: Energy Nexus Group. 2002. *Technology characterization: microturbines.* U.S. Environmental Protection Agency. Washington, D.C.)

---

[*] For example, Kawasaki offers a version of their M1A 13X, 1.4-MW gas turbine with a catalytic combustor with less than 3 ppm NO_x guaranteed.

Microturbine Emissions Characteristics

Table 10-4 presents typical emissions for microturbine systems. The data shown reflect manufacturers' guaranteed levels.

CHP is a well-accepted technology and there may be a variety of federal and state funding as well as financial and technical expertise available to assist with a new or modified project.

CHP FUNDING OPPORTUNITIES*

CHP funding opportunities are offered by various entities throughout the country, many at the state and federal level. These opportunities take a variety of forms, including the following:

- Financial incentives, such as grants, tax incentives, and low-interest loans
- Regulatory treatment, such as expedited permitting, recognition of environmental benefits, and simplified interconnection

As a service to their partners, the EPA CHP Partnership provides a listing of state and federal CHP incentives. This information is reviewed and updated twice a month and is available online. At the time of writing, this URL was www.epa.gov/CHP/funding_opps-bio-htm.

- Alaska Energy Authority—Alaska Power Project Loan Fund
- Arizona Department of Commerce—Energy Office—Arizona Municipal Energy Management Program
- Bonneville Environmental Foundation—BEF Renewable Energy Grant
- California Energy Commission—California Emerging Renewables Rebate Program, California Supplemental Energy Payments (SEPs)
- California Public Utilities Commission—California Self-Generation Program, California Utility Rates to Support Combined Heat and Power
- Connecticut Clean Energy Fund—Connecticut Property Tax Exemption
- Connecticut Office of Policy and Management—Connecticut New Energy Technology Program
- Idaho Department of Water Resources—Energy Division—Idaho Low Interest Energy Loans
- Idaho Energy Resources Authority—Idaho Renewable Energy Generation Benefit
- Illinois Clean Energy Community Foundation—Illinois Clean Energy Community Foundation Grants
- Indiana Department of Commerce, Energy and Recycling Office—Indiana Alternative Power and Energy Grant Program—Indian Distributed Generation Grant Program (DGGP)
- Indiana Department of Environmental Management—Indiana NO_x Budget Trading Program Energy Efficiency
- Iowa Department of Natural Resources—Iowa Methane Gas Conversion Property Tax Exemption
- Iowa Energy Center—Iowa Alternate Energy Revolving Loan Program
- Kansas Corporation Commission Energy Office—State Energy Program Grants
- Massachusetts Department of Environmental Protection—Massachusetts NO_x Budget Trading Program Energy Efficiency

---

* Source: The U.S. Environmental Protection Agency (EPA).

- Massachusetts Technology Collaborative—Massachusetts Com/Ind/Inst DG Initiative Grants
- Minnesota Department of Commerce State Energy Office—Minnesota Anaerobic Digestion Production Incentive
- Mississippi Development Authority—Energy Division—Mississippi Energy Investment Program
- New Mexico Energy, Minerals and Natural Resources Department—New Mexico Industrial Revenue Bond (IRB) Financing, New Mexico Public Facility Energy Efficiency Loan, New Mexico Renewable Energy Production Tax Credit
- New York State Department of Environmental Conservation—New York NO_x Budget Trading Program Energy Efficiency
- North Carolina Solar Center—North Carolina Renewable Energy Tax Credit
- North Carolina State Energy Office—North Carolina Energy Improvement Loan Program
- Ohio Department of Development Office of Energy Efficiency—Conversion Facilities Property Tax Exemption, OH Energy Efficiency Grant and Loan Program
- Oregon Department of Energy—Oregon Small Scale Energy Loan Program (SELP), Oregon Business Energy Tax Credit
- Pennsylvania Department of Environmental Protection—Pennsylvania Energy Development Authority
- South Carolina Energy Office—South Carolina ConserFund Loan Program
- South Dakota Public Utilities Commission—South Dakota Local Property Tax Exemption
- State Technologies Advancement Collaborative—Energy Efficiency Research and Development, Demo, and Rebuild America Project
- Tennessee Department of Economic and Community Development—Energy Division—Tennessee Small Business Energy Loan Program
- Utah Energy Office—Utah Energy Systems Tax Credit

The EPA also has a "brownie points"/good citizen award program related to CHP activity. This information is available at: http://www.epa.gov/chp/partner_resources/pub_recog.htm. (If this link does not work, visit http://www.epa.gov and browse the EPA's website.)

ECONOMICS OF CHP*

The following work was done by Oak Ridge National Laboratory, who works for government environmental agencies on alternative energy research. Their market assessment of all CHP methods available is included here, so that a comparison of microturbine CHP systems versus other CHP systems can be made.

CHP has the potential to dramatically reduce industrial-sector carbon and air pollutant emissions and increase source energy efficiency. Industrial applications of CHP have been around for decades, producing electricity and by-product thermal energy on-site, and converting 80% or more of the input fuel into useable energy. Typically, CHP systems operate by generating hot water or steam from the recovered waste heat and using it for process heating, but it also can be directed to an absorption chiller where it can provide process or space cooling. These applications are also known as cooling, heating, and power.

---

* Source: Resource Dynamics Corporation. 2003. *Cooling, heating and power for industry: a market assessment.* Oak Ridge National Laboratory. Oak Ridge, TN.

The focus of this study was to assess the market for cooling, heating, and power applications in the industrial sector. New thermally-driven cooling technologies are being developed and demonstrated that can potentially utilize the CHP heat output effectively for uses typically reserved for high-value electricity. CHP with cooling has potential applications that could not be economically served by CHP alone, creating an additional use of the by-product thermal energy when heating loads are minimal. Now more than ever, CHP can potentially decrease carbon and air pollutant emissions while improving energy efficiency in the industrial sector.

Today there are a variety of cooling technology options for cooling, heating, and power. Absorption chillers are available that can be installed with a CHP system to utilize the heat output to produce process cooling. For the purpose of this study, engine-driven chillers are considered cooling, heating, and power applications since they provide thermal output, in the form of process or space cooling, and also displace electricity that typically would be purchased to provide this cooling. Desiccant technology, which can be used in conjunction with CHP waste heat to provide cooling, has not been considered due to deficiencies in energy-use data but would add to the market potential estimated in this effort.

The focus of this study was on smaller CHP technologies, otherwise known as distributed generation (DG). *DG* is defined here as power generation smaller than 50 MW with the unit output being used either on-site or close to where it is produced. Other potential uses of DG technologies include peak shaving, premium power, and "green" power. However, given the study objective of assessing the most likely markets for CHP technology, the focus of this study was on cooling, heating, and power applications, with straight power generation (i.e., without heat recovery) also included.

To determine the potential for cooling, heating, and power in the U.S. industrial sector, this effort evaluated a wide range of DG units. The study focused on units due for production by year 2002 (base case scenario) and includes reciprocating engines, industrial turbines, microturbines, combined-cycle turbines, and phosphoric acid fuel cells. Table 10-5 summarizes the scope of this effort. A future case is included as a sensitivity and considers significant improvements in cost and performance for each of these technologies, as well as the emergence of solid oxide fuel cells.

Market Potential and Market Penetration

The market potential was estimated for cooling, heating, and power applications, including power generation without heat recovery, straight CHP, CHP with absorption cooling, and engine-driven chillers for process cooling. As shown in Figure 10-9, the potential for these applications in the U.S. industrial sector is estimated at 33 GW of power-generating capacity with currently available technology.

In Figure 10-9, market estimates show that almost three quarters (about 24 GW) of the current potential is for straight CHP applications, where the waste heat from power generation is used for process heating. CHP with an absorber represents 15% of the potential (about 5 GW), serving industries with substantial cooling demand, including the chemical and petroleum industries.

There is still a market for straight power generation, without recovering the waste heat, mostly for larger industrials that can accommodate larger (20 to 50 MW) combined cycle units that offer low cost of generation. These applications are strong in California, with its high grid prices, as well as with large primary metals facilities with limited steam/hot water demands.

There is little potential for engine-driven chillers based on their power-generating capacity, but they represent over a third of the CHP cooling capacity on a tonnage basis

TABLE 10-5   Summary of Study Scope

	Size (MW)	*Applications*	*Technologies*
Included	Up to 50 MW	Combined heat and power (CHP) Cooling, heating, and power Straight power generation (no heat recovery)	Reciprocating engines Microturbines Industrial turbines Combined-cycle turbines Fuel cells Absorption chillers
Not included	Greater than 50 MW	Peak shaving Backup/emergency power "Green" power	Renewables Desiccants

Source: Resource Dynamics Corporation. 2003. *Cooling, heating and power for industry: a market assessment*. Oak Ridge National Laboratories. Oak Ridge, TN.

in this analysis (see Figure 10-13 for more detail). The details of the analysis provide some insight into this market breakdown when aspects such as the type of technology, regional characteristics, and industry needs are considered.

Based on data from the Energy Information Administration, it is estimated that current CHP use in the industrial sector for units under 50 MW is about 11 GW. Comparing this value with the market potential estimate of 33 GW, it would appear that the market penetration is about one third. While there are few data on penetration rates of new CHP technology under current economic conditions, the commercial and industrial decision-making process regarding energy-related investments in general can yield some insights.

Many studies on the acceptance of energy-saving investments examine the amount of time necessary to pay back the investment. While the market potential presented here is based on a 10-year cash flow analysis to determine the option with the best net present value, a simple payback was also calculated.

Figure 10-10 illustrates for the current case that almost 20% of the applications offer a payback in under 2 years, and almost 60% (about 20 GW) of the applications deemed economically feasible have a payback in under 4 years. Assuming that the 11 GW that has been installed was taken from the more attractive paybacks, that means that about

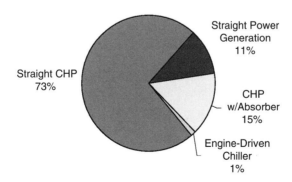

FIGURE 10-9   *U.S. industrial cooling, heating, and power market potential (33 GW). (Source: Resource Dynamics Corporation. 2003.* Cooling, heating and power for industry: a market assessment. *Oak Ridge National Laboratory. Oak Ridge, TN.)*

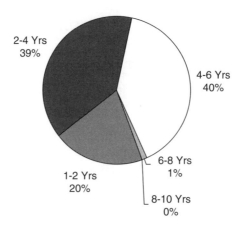

2-4 Yrs
39%

4-6 Yrs
40%

6-8 Yrs
1%

1-2 Yrs
20%

8-10 Yrs
0%

FIGURE 10-10   *Distribution of payback periods for potential industrial combined heat and power applications (33 GW). (Source: Resource Dynamics Corporation. 2003.* Cooling, heating and power for industry: a market assessment. *Oak Ridge National Laboratory. Oak Ridge, TN.)*

9 GW of under 4-year payback market potential is still unrealized. This portion is likely impeded by market or regulatory barriers discussed in Section 4 of the complete Oak Ridge National Laboratory report.

A general rule of thumb is that a 2- to 4-year payback is required for industrial facilities to purchase equipment that will reduce their energy bill. With 9 GW of applications offering paybacks in this range that have not been installed, the indication is that further cost reductions or economic assistance (e.g., tax incentives or rebates) may be required to stimulate this attractive but unrealized portion of the market.

Market Potential by Technology and Size Range

Figure 10-11 illustrates the projected market potential by CHP technology. The market analysis shows that in the base case, the CHP marketplace is shared by turbines and reciprocating engines.

In the future, turbines are projected to adopt many of the high-efficiency features pioneered by the DOE Advanced Turbine Systems (ATS) Program, and thus take CHP market share from ATS, combined cycle systems, and even improved reciprocating engines. The reason for the turbines capturing future market potential is that the electrical efficiency improvements projected for turbines are much greater (relative to their current efficiency) than those projected for reciprocating engines, while turbines continue to hold an advantage in terms of quality of thermal output.

In the future scenario, overall market potential improves as well, almost reaching 50 GW (from 33 GW). Although not shown in Figure 10-11, in the future, fuel cells are projected to drop below $900 to $1200/kW installed with the development of molten carbonate and solid oxide technologies, and emerge in the future CHP marketplace with less than 5 MW of capacity. This penetration could continue if further improvements in fuel cell cost (i.e., below $1200/kW) are attained.

Figure 10-12 shows how the market potential of CHP technologies varies by the size of the generating unit. In the base case, engines dominate in the smaller sizes (under

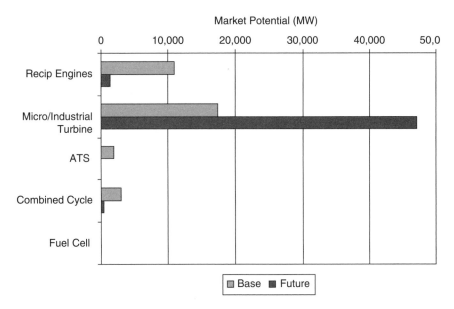

FIGURE 10-11   *Market potential by technology (MW). ATS = Advanced Turbine Systems. (Source: Resource Dynamics Corporation. 2003. Cooling, heating and power for industry: a market assessment. Oak Ridge National Laboratory. Oak Ridge, TN.)*

1 MW) over microturbines and fuel cells. Their combination of high efficiencies and competitive installed cost makes them hard to beat. In the mid range (1 to 20 MW), turbines take over, due to the large concentration of CHP-compatible sites in this size range.

Turbines offer better economics for CHP when most or all of the thermal output is valued. In the larger sizes (20 to 50 MW), turbines do well in CHP applications and combined cycles emerge, offering economic potential for baseload power applications. The combined cycle applications are attractive in industries (such as steel) with relatively low steam demands and for larger plants in states with high retail rates, such as California.

In the future, the turbine CHP market potential greatly expands as microturbines take over in the under-1-MW applications, and larger (over 1 MW) turbines benefit from improved electrical efficiency and lower capital-cost-per-unit power output. Again, many of these improvements are seen as resulting from the ATS program for over-1-MW turbines, and similar improvements in microturbines and engines are expected to result from the DOE's Advanced Microturbine and Advanced Reciprocating Engine Systems programs.

Furthermore, as previously shown in Figure 10-9, about 16% of the potential applications of CHP favored the generation of cooling from the CHP unit. Four different cooling operating strategies were explored, including single effect absorption units and engine-driven chillers, both baseloaded and serving the entire cooling load. The market potential, in terms of cooling tons, is shown in Figure 10-13 for each of these four strategies.

Figure 10-13 shows that engine-driven chillers are competitive in the smaller size ranges, particularly for serving the entire cooling load (sized to peak). In the 10- to 50-ton range, engine-driven chillers sized to peak offer the potential for over 140,000 tons of cooling. Peak-sized absorbers also do well in this smaller range, representing over

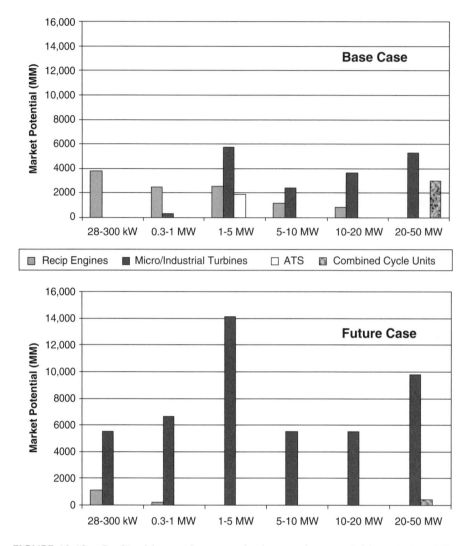

FIGURE 10-12   *Combined heat and power technology market potential by unit size. ATS =*
Advanced Turbine Systems. (Source: Resource Dynamics Corporation. 2003. Cooling, heating
and power for industry: a market assessment. *Oak Ridge National Laboratory. Oak Ridge, TN.)*

100,000 tons of potential cooling. A similar but lower potential is demonstrated in the
50- to 100-ton range, and as the on-site cooling load grows, the potential for baseload
absorbers takes hold. This technology and operating strategy leads the remainder of the
cooling size ranges, topped off by the 1000- to 2000-ton range, where baseload absorbers
show most of their potential. In this size, the capital cost of absorbers drops significantly,
and the economics improve as a result.

Four scenarios were constructed to evaluate how sensitive the base case is to varying
inputs. In the first scenarios (the Future Case), the focus was on how improvements in
CHP cost and/or efficiency impact potential market size. The remaining three sensitivities
were added to illustrate the effects of changing natural gas prices on the CHP market for
industrial applications, including the accompanying effect on retail electric prices.

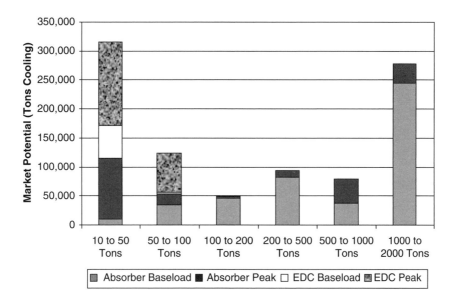

FIGURE 10-13 *Combined heat and power cooling market potential by range of cooling unit size. EDC = engine-driven chillers. (Source: Resource Dynamics Corporation. 2003.* Cooling, heating and power for industry: a market assessment. *Oak Ridge National Laboratory. Oak Ridge, TN.)*

A high price scenario (High) reflected a jump in gas prices that remained high throughout the life of the unit. The moderate price scenario (Moderate) reflected a temporary jump followed by a return to lower prices. In both of these cases, retail electricity prices (the energy component) were adjusted upward to reflect the impact of the higher gas prices on wholesale electric prices, using a methodology similar to how utilities calculate their fuel-adjustment clauses.

The Peak scenario uses the high gas price scenario but reflects the gas price impact on the demand component of the retail electricity prices. Appendix A provides more detail on these scenarios.

Overall market potential results of the sensitivity analysis (Figure 10-14) indicate that improvements in installed cost and efficiency increase the potential market size dramatically. The Future Case (reflecting improved CHP cost and performance) increases the potential market from 33 to almost 50 GW, a 50% increase in the market size.

The impact of increasing natural gas prices is shown in the Moderate, High, and Peak scenarios as decreasing the market potential for industrial CHP, even with accompanying increases in electric prices. As shown in Section 3 of the complete Oak Ridge National Laboratory report, the decrease in market potential was lower in regions where concentrations of natural gas generation exist in the wholesale electric market, including West South Central, Pacific, and parts of the Northeast and Mid-Atlantic. This effect will likely diminish somewhat as more gas-fired generation is placed in service in these regions to meet future capacity needs.

Despite improving economics, increasing emphasis on overall energy efficiency, and concerns over restructuring of the electric utility industry, CHP systems face challenges for further penetration in the industrial market. The realization of benefits inherent in implementing CHP on a wide scale is hindered by a combination of barriers in the following categories:

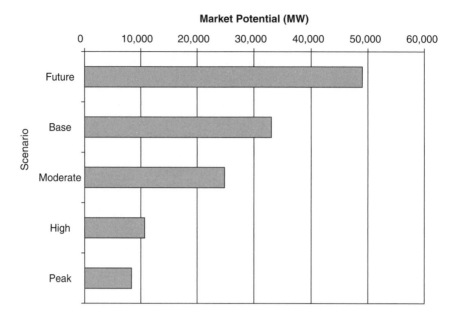

FIGURE 10-14   *Future scenario offers highest market potential. (Source: Resource Dynamics Corporation. 2003.* Cooling, heating and power for industry: a market assessment. *Oak Ridge National Laboratory. Oak Ridge, TN.)*

- Economics and tax treatment
- Product performance and availability
- Awareness, information, and education
- Utility policies and regulation
- Planning, zoning, and codes
- Environmental regulation
- Supporting market infrastructure

These barriers can often make a CHP project uneconomical and can frequently present a confused and uncertain option to potential end users.

To overcome these barriers and maximize the many benefits of industrial CHP, further research and development is needed to allow these technologies to compete with more conventional options. The CHP and thermal cooling technologies markets both need lower costs, increased efficiency, reduced maintenance, greater reliability, and lower emissions.

Chapter 11

Unconventional
Microturbine Fuels

Microturbines are smaller-sized gas turbines, so for the most part, they can use the same growing range of fuels that gas turbines can use. In the last decade, successful pilot plants and even established nonprototype installations have burned fuels that include, but are not limited to, the following:

- Waste liquor from paper manufacture
- Biomass
- Pulverized coal
- Waste fluids that are by-products from petrochemical and plastics manufacture
- Low-BTU gas or fluid that is some process plant by-product
- Landfill gas (LFG)
- "Swamp" gas from marshes

Many of these "waste" fluids have a lower BTU value than clean natural gas. Gas turbine manufacturers (whether their units are "standard size" or microturbines) have to address issues including the following:

- Adequate combustion
- Tolerable levels of carbon monoxide and unburned hydrocarbon in the waste gases
- Removal of acrid or acid-forming by-products to the extent that recuperator surfaces or waste heat recovery units (that are part of combined heat and power [CHP] systems, among others) are not eaten away by, for instance, by-products of sulfur

If their fluid is lower in BTU than methane gas, they may be able to solve some of their design problems with longer residence times in the combustor.

Many waste and rotting-waste processes produce gas that contains methane and other hydrocarbons to some extent. These processes are found in, but not limited to:

- Landfills
- Rotting dung (i.e., cow manure, human waste, other animal waste)
- Swamps

The quality of the evolving gas will vary depending on the following:

- The nature of the waste
- The surrounding climate
- Moisture content
- A series of other factors

It is possible to collect these gases using a network of vertical and horizontal channels and pipes, with appropriate collecting points, valves, and conduits. The composition of the gas or fluid and the quantity available often make the decision about what item of power-generation machinery can be used to produce power, using it as fuel.

Other factors dictating machinery choice are the same ones as affect the purchase of a large industrial-sized gas turbine or gas turbine combined cycle. These include but are not limited to the following:

- Grants, federal or state, that are available
- Size of the power-generation unit
- Spares and service available for the power-generation machine(s) of choice
- Known end-user experience on costs of ownership
- Reliability and availability figures
- Financing available
- Skill levels of operations personnel available for hire
- Overall demographics including geographical location, climatic conditions, and proximity to required supplies

So, if one considers all of the above, one may conclude, depending on the circumstances, that the following are true:

- Although a reciprocating engine generating set is a well-known, "tried and true" workhorse with better reliability and cost of ownership per fired hour per kilowatt (kW) of power, the smallest-sized unit one can get is too large for the power desired or the fuel available, or both.
- Although a fuel cell appears to be an ideal item of equipment with its clean burn of hydrogen (that it produces from the fuel), if sulfur is inherent somewhere in the system or fuel, hydrogen sulfide (H_2S) issues may surface, and the stack may get eaten away by acid that the H_2S forms.
- Microturbine(s) may be the only fuel driven equipment available in small enough kW sizes and easy enough to install at various locations of, for instance, a large landfill.
- If it is possible to use the microturbines in CHP application, then the sponsors may qualify for all kinds of grants, government-awarded public relations certificates, and financing.
- The legislation to which one is subject, may be a deciding factor. In California, one gets both federal and state rebates for certain kinds of alternative energy.
- The available backup sources of power (e.g., diesel genset, local grid) may be a deciding factor.
- Tariffs for local grid electricity and tax structure may dictate that, although the power produced is more expensive than available grid power, rebates and incentives still make it worthwhile to continue being a small or independent power producer.
- Projected electricity tariffs and/or projected natural gas and oil prices, as well as political issues with international fossil fuel supply chains, may also force what may look like a project that will lose money for its first years to go ahead anyway.
- Certain inappropriate choices (e.g., in Case 11-1 below, a research team in Florida, picked a fuel gas booster compressor supplier in England) can result in downtime while spare parts are awaited.
- The gas quality sought may require complex pretreatment that changes the economics of a case (see Case 11-1 below).
- The support hardware involved (e.g., piping, joints) may prompt periods of downtime, if the system as a whole is intolerant of moisture infiltration or gas condensation (see Case 11-1 below).

- One needs to predict gas fuel flow available throughout the intended operating cycle and control this gas flow appropriately for the demand load. If this is not done adequately, a microturbine, particularly a finicky one, will experience downtime (see Case 11-1 below).

The question of picking an operating system is therefore a complex one. However, certain government bodies are tasked with exploring the permutations and combinations of various types of equipment choice and system design. Potential end users ought to first explore these project reports before proceeding. Frequently, the government bodies in question place them on the Internet. However, the end user needs to check if the report does n ot have update information. Generally, government bodies are good about providing a forwarding e-mail link if they change Internet addresses.

The following case studies are varied in nature. Some are quite specific, like the one that uses cow manure to produce fuel gas. Others, like the California Public Service Commission study, explore many types of power plant options. Their work is interesting because it offers readers comparisons of how different power-generation plants fared within the California infrastructure. The study by Natural Resources Canada deals with a project conducted in California, but its data are presented differently than with the former. For each case, the original material has been edited to varying extents. The sources can all be contacted via their current contact information, affiliation head offices, or Internet addresses.

In brief, the treatments below describe the following cases:

- Case 11-1: The results from an LFG-burning microturbine project in Florida were erratic, perhaps due to the fact that it was, in part, an academic project. The project achieved some noteworthy results and was supported by government funding. Part of the paper is paraphrased here and included as a good example of government proactive participation in student education.
- Case 11-2: This case, involving fuel gas from cow manure, is interesting for several reasons. To start with, it demonstrates the potential for using farm animal dung as a fuel. It also shows a closer partnership, as well as a more modest budget (compared with grants in some richer states) afforded by the Iowa Department of Natural Resources (IDNR). This case is interesting in that the participants are few in numbers but resourceful and determined. The author considered leaving out much of the detail in this rather thorough report, but has not done so. There are a great many farm animal complexes globally that could emulate this example.
- Case 11-3: Extracts from "Economic and Financial Aspects of LFG to Energy Project Development in California" summarize an excellent and thorough study done by the California Energy Commission that demonstrates how the microturbine fares in competition with other energy sources.

CASE 11-1: MICROTURBINE GENERATOR RUNNING ON LANDFILL GAS IN FLORIDA*

The performance of a low-emission microturbine generator producing power from LFG was investigated. Previously, methane gas emitted from the garbage was pumped out and burned off using a landfill flare stack. This method releases toxic fumes into the

---

* Taken from: Stefankos, E. K. et al. 2004. *Landfill low emission micro turbine generator*. University of South Florida. Tampa, FL.

atmosphere and wastes energy. The microturbine generator was intended to reduce greenhouse gases (GHGs) while generating grid-connected power.

To use methane gas from the landfill, processing to put it into proper condition for energy conversion was required. For processing, a series of filters and a compressor running off the main pumping station at the Hillsborough Landfill test site in Tampa, FL, were used.

The impurities were removed and the gas brought up to appropriate delivery pressure for a Capstone low-emission microturbine generator. The electric power produced was provided to the TECO electric grid. Typically, the LFG ratio is approximately 60% methane and 40% carbon dioxide (CO_2).

Experimental Setup

The process of generating power from the methane gas was designed to attempt maximized output. The system (Figure 11-1) shows the devices used in the process. The system uses the existing landfill pump station and LFG supply pipeline.

The gas was pumped into piping leading to the desiccant dryer, which takes out most of the moisture and sends it to the condenser sump. Because of the way biogas is produced, it is saturated with water. Any cooling of the gas in the process lines almost always generates a liquid condensate when the gas first enters the system. The gas was then pumped into the rotary compressor that brings the pressure up to around 80 to 90 pounds per square inch (psi). The pressurized gas is then piped to the refrigerated dryer that removes more moisture from the biogas. Removal of the condensate, followed by heating the gas, produces a dry gas with a temperature above its dew point. Figure 11-2 shows a flowchart of how the turbine system at the site is arranged.

The gas was pumped into the siloxane filter. Siloxanes are composed of carbon, hydrogen, oxygen, and silicon and must be removed to prevent damage to the turbine. Tiny particles of silica form in the combustion section when siloxanes are present in the fuel to the microturbine. Over time, the abrasive silica particles cause erosion of some of the metal surfaces they contact.

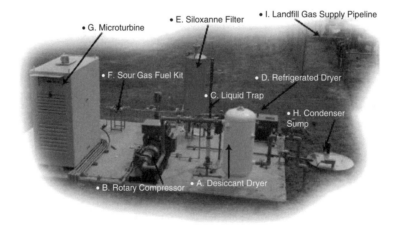

FIGURE 11-1   *Methane microturbine generator. (Source: Stefankos, E. K. et al. 2004.* Landfill low emission micro turbine generator. *University of South Florida. Tampa, FL.)*

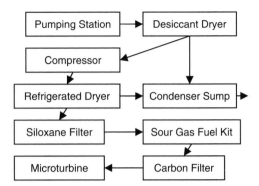

FIGURE 11-2 *Diagram showing flowchart of microturbine system. (Source: Stefankos, E. K. et al. 2004.* Landfill low emission micro turbine generator. *University of South Florida. Tampa, FL.)*

Spare filtration capacity ensures that the methane is dry, clean, and at least 18°F (10°C) above its dew point, allowing for both the hottest and the coldest ambient temperatures.

Then the gas flows through a sour gas fuel kit and a carbon filter. The microturbine flame design temperature is 1100°F.

The compressor requires 4 kW to keep the pressure constant in the system and receives this power from the grid. The microturbine continues to be fed methane gas as long as the flare stack burns. The Hillsborough pumping station shuts down if it detects that the flare has gone out.

Results

The power output as shown in Figure 11-3 was highly variable. Reasons for the variation include the following:

1. Starting the microturbine
2. Errors in sizing the methane fuel piping
3. Compressor failure
4. Air contamination of the methane fuel
5. "Sticky" microturbine fuel control valve

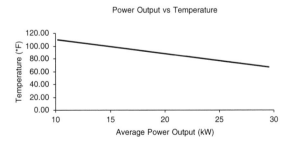

FIGURE 11-3 *Average power output at compressor inlet temperature. (Source: Stefankos, E. K. et al. 2004.* Landfill low emission micro turbine generator. *University of South Florida. Tampa, FL.)*

In comparison to the emissions from the flare stack, the microturbine creates almost a negligible amount of GHGs. The data, shown in Figure 11-4, more clearly demonstrate the significant difference.

The data are based on 8760 hours per year. The microturbine emits 0.006617 tons of nitrogen oxides per year as opposed to the calculated 2.45 tons of nitrogen oxides per year emitted by the flare. The results clearly show that the microturbine emits significantly less nitrogen oxides than the flare. This demonstrates another positive aspect of the microturbine in its current implementation.

The turbine has been producing between 24 and 30 kW of output power since its initial start-up (see Figure 11-3).

This turbine was also affected by fuel pressure. When temperatures were the lowest, the methane settled and caused the purity of the gas to decrease. This caused several power outages.

During the summer months, when temperatures reached the 100°F range, the turbine would shut down frequently. This was resolved by replacing the existing ½-inch tubing with 1-inch tubing. This allowed more gas through the piping and also lowered the chance of debris being stuck in the piping. Pressure loss across the entire system decreased from 20 to 7 psi. The compressor did not have to work as hard to maintain a constant pressure. The turbine typically needs 75 psi to run. By replacing the tubing, the pressure was increased from about 66 to 86 psi. The methane percent content in the gas increased from 32% to about 36%.

The turbine was inoperable during the months of July and August due to compressor failure. The compressor was under warranty, but the company had to build and ship a new compressor from England. The new compressor ran more smoothly and, most importantly, stayed in an idle position when it was not needed.

Figure 11-5 shows the net energy that was produced during the course of the research for each month. As shown in this graph, during the months of June through August, the turbine was nonoperational. The total kW hours produced during the year is shown in Table 11-1.

CASE 11-2: FUEL GAS FROM COW MANURE*

Making electricity from manure on a 700-cow dairy farm started with a grant proposal to the IDNR. The writer, R. Crammond, proposed a digester similar to one he

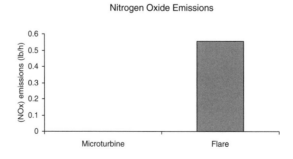

FIGURE 11-4   *Nitrogen oxide emissions (lb/hr). (Source: Stefankos, E. K. et al. 2004.* Landfill low emission micro turbine generator. *University of South Florida. Tampa, FL.)*

---

* Source: Meyer, D. Presented at the BioCycle Conference on Renewable Energy from Organics Recycling.

Average Power Output Per Month

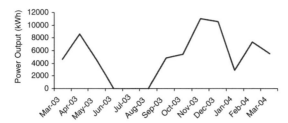

FIGURE 11-5   *Net kilowatt hours per month. (Source: Stefankos, E. K. et al. 2004.* Landfill low emission micro turbine generator. *University of South Florida. Tampa, FL.)*

designed in Michigan—a plug flow design with a concrete lid. The grant application was successful.

Constructing the Digester

Before the anaerobic digester became a reality, Top Deck (the grant recipient) had been planning an expansion of the farm that would more than double the herd size from 300 to 700 cows. One of the IDNR's requirements was that the farm had to complete expansion before installing the digester. As the first item of business, Top Deck had to modify existing earthen manure storage of 180 × 240 feet and 10 feet deep to meet the latest requirements. This took several months.

TABLE 11-1   Energy Output During Test Year

Month of Production	Total Power Output (kWh)
March 2003	4673
April 2003	8654
May 2003	4446
June 2003	0
July 2003	0
August 2003	0
September 2003	4884
October 2003	5373
November 2003	11,045
December 2003	10,586
January 2004	2882
February 2004	7321
March 2004	5485

(Source: Stefankos, E. K. et al. 2004. *Landfill low emission micro turbine generator.* University of South Florida. Tampa, FL.)

The digester project started with construction of the 27 × 124 × 12 feet concrete digester tank. A 4-inch drainage pipe was placed around the perimeter of the tank (insulated with polystyrene foam). Winter temporarily halted the work. Construction restarted in the summer of 2001 when the precast lid was installed. The 4-inch-thick lid was covered with 4 inches of foam insulation and 1 foot of soil.

Two additional tanks were added at the west end of the digester—a preheat tank and a separator pump tank, both 13 × 13 × 12 feet. The precast concrete lids for the two smaller tanks were each cast in two sections that were 6½ × 13 feet. This allows the lids to be removed more easily to provide access to the interior equipment (e.g., the heating pipe grid in the preheat tank). Together the two tanks plus the digester provide about a 14-day detention time based on 29 gallons of manure per cow per day, plus the milking center wastewater.

The first tank preheats 20,000 gallons of daily manure production to 98°F. The preheated manure is pumped from this tank to the east end of the digester through an 8-inch polyvinyl chloride (PVC) pipe. The preheat tank was included for two reasons: it provides much easier access to the heat exchanger for maintenance or removal, and it can serve as a settling tank for solids from the dairy, such as sand and grit. It is preferable to have these solids settle in the tank rather than in the digester.

The other small tank was installed to house, in the future, a pump to transfer the digested manure to a separator that can capture the manure solids for use as bedding for the cows' free stalls. The separator was not included initially because it was preferred to get the digester system up and running first. Initially, the manure from the digester flowed over the wood weir, entered the separator pit, and then flowed out by gravity through a 12-inch pipe to the earthen manure storage lagoon. When a mechanical separator is installed, it can be located just west of the current pit.

The IDNR wanted a surface outlet for the digester manure. This necessitated constructing the digester partially above ground, with the top 6 feet above the existing grade. This avoided the need for a pump to empty the effluent into the earthen storage, which is located about 30 feet to the south. It also allowed any gas leakage from the digester gas to remain above ground level. Underground leaks have the potential to migrate horizontally and cause an explosion elsewhere. Therefore, the engine/generator building was located 40 feet from the digester.

Power System Layout

In the spring of 2001, Alliant Energy Corporation, in exchange for the electricity produced on the farm, offered to furnish the electricity-generating equipment. The equipment purchased includes a Waukesha 150-horsepower (HP) engine (used) and a 100-kW synchronous parallel generator that runs at 1200 revolutions per minute (RPM). A 30-kW Capstone microturbine generator was also included in the system. The microturbine incorporates a compressor, recuperator, combustion chamber turbine, and permanent magnet generator. These rotating components are mounted together on a single shaft supported by air bearings that rotate at up to 96,000 RPM. The microturbine is air cooled and thus eliminates the need for a liquid coolant. Gas from the digester needs to be cleaned, compressed, and dried before entering the microturbine combustion chamber.

The contract to supply the generating equipment and controls was awarded in March 2001 to Perennial Energy of West Plains, MO. The Waukesha engine and generator and much of the microturbine arrived in mid December on an assembled skid packaged. The remaining microturbine hardware, an oil bath system for drying the gas, was delivered in April 2002. The generating equipment can be purchased by Top Deck Holsteins at the end of the 10-year contract for a 10% salvage fee. The contract with Alliant also compensates Top Deck $500 per month (adjustable with kW hours generated) as a maintenance fee for monitoring the digester and the engine/microturbine equipment.

The building that houses the generation equipment is a 24 × 36–foot insulated wood-framed shed with 13-foot-high walls. The interior walls are finished with drywall for sound insulation. The building includes a gas space heater plus two 36-inch-diameter ventilation fans on one side wall and four 36-inch-square louvers on the opposite wall. There is one water drain required by the microturbine to handle the moisture removed from the gas.

Heat for Digester and Dairy

Heat captured from the engine-generator and microturbine system is used to maintain the digester temperature and supply heat to the dairy center. When hot water from the generation equipment is insufficient (such as during startup), boilers running on lique-fied petroleum (LP) gas are used as heat sources.

To heat manure for digestion, hot water is circulated through separate steel pipe heat exchangers installed in the digester and the preheat tank. The digester contains 290 feet of 2-inch pipe that is 18 inches above the floor and 12 inches in from the outer wall. In the preheat tank, the 1000 feet of 2-inch pipe is arranged in a grid loop that measures 7 feet high, 6 feet wide, and 11 feet long. The grid starts 2 feet above the tank floor. Manure constantly covers this pipe grid to prevent pipe rusting. Each heat exchanger has a single 1-HP pump to circulate water.

A third hot water loop runs 370 feet from the engine/generator building to the dairy center. This hot water loop feeds a heat exchanger inside a hot water heater in the dairy. The hot water from the engine/generator runs through a heat exchanger (rather than used directly) because it contains antifreeze (needed when the system is inactive). The energy from this hot water is used at the dairy center in many ways: for cleaning milk pipelines; for space heating in the milking parlor via a radiator and fan; for space heating via pipes embedded in a floor; and for high-pressure hot water washing in the milking parlor.

Hot water is captured from the engine and microturbine as the heat becomes available. The generation system's computerized controls determine when and where to pump water based on temperature set points. A total of 460,000 BTU/hr is provided by the engine water jacket and the engine exhaust pipe heat exchanger at full capacity. Another 270,000 BTU/hr is available from the microturbine exhaust pipe heat exchanger when running at full RPM.

When the engine is running below full load, there may not be enough hot water to maintain the digester temperature. To sufficiently preheat manure for digestion manure requires about 343,000 BTU/hr in cold weather (based on a 40°F manure temperature and a 98°F digester temperature). When little or no heat is available from the generation equipment, this energy must come from fuel combustion. So two 200,000-BTU/hr boilers were rented that together use about 70 gallons per day of LP gas. In hindsight, the boilers should not have been rented. It took an unexpectedly long time to get the digester up and running perfectly. A 400,000-BTU/hr boiler could have been purchased for the cost of renting them off and on for 5 months. The boiler would be useful when both the engine and microturbine go down for repairs. Fortunately, in normal situations only one is down for repairs at a time, so the other one can heat the hot water.

Making Adjustments

Several problems occurred that required system changes, especially during commissioning. Also, experience with the digestion suggests other changes that promise to improve its operation.

Among the first problems evident was a lack of pressure inside the digester due to poor sealing. Based on the experience at another anaerobic digestion project, a new polyurea spray sealant was used to seal the digester instead of the more conventional epoxy tar. The spray polyurea was used initially at a 100-ml thickness on the inside of the digester at the crack between the wall and lid and also on both sides of the "gas beam" (located in front of the overflow weir to hold the gas in the digestion tank). Because H_2S in the biogas deteriorates concrete, the spray product was also used to coat the tank surfaces exposed to biogas (lid area and down 3 feet from the lid on the walls). The product, which cost about $4/square foot, was supposed to resist up to 85% pure sulfuric acid and handle 520% elongation (it stretches with temperature). However, the digester failed to hold pressure. The remedy was to bring back the sealer company and seal the outside edge of the tank where the lid sits on the wall.

Another problem that developed was that foam and liquid rose up into the 3-inch biogas line leading from the digester to the engine/generator system. This caused plugging problems in the 3-inch biogas line and engine fouling. The remedy was to install a 30-gallon vessel with a drain and an 8-inch diameter window to check for liquid and foam in the biogas line. At the same time, a water spray nozzle was added to spray down the foam at the end of the biogas line just upstream of the 30-gallon tank.

A flare was added to burn the gas produced during turbine downtime. Other items added included a gas meter and a 100% shutoff valve outside the building.

Effects of Bedding

The farm's bedding preferences influence the digester operation. Currently, the farm uses rice hulls for bedding. If the rice hulls do not decompose well enough in the digester, they will accumulate in the tank. Eventually, the farm may have to switch to another bedding material. However, in the meantime, it will be necessary to occasionally clean out sediment in the digester tank.

Fortunately, there are five 10-inch diameter perimeter pipes on the north side of the digester that were placed at a 60-degree angle in case the digester needs to be pumped empty. Also, in the event that a person must enter the digester pit for clean-out, the 10-inch PVC pipes can be used for ventilation by inserting a small fan in a pipe. There is also a 12-inch valve that can be opened to drain the digester into the 13 × 13–foot preheat tank, where there is a 20-HP manure pump (with a propeller) to remove manure if necessary. Also, there is a 14 × 14–inch opening in the common wall between the preheat tank and the separator pit. If the preheat tank is ever overloaded, manure will discharge by gravity into the separator pit and then flow into the storage lagoon.

Simplified Economics and System Performance

Manure is pumped independently from the North and the South barns, which each house about 350 cows, to the preheat pit at the end of the digester. The North barn's manure has the milking center wastewater added to it. Located in the pit, a 20-HP Houle pump with a propeller agitates and pumps the manure into the digester. The pump operates 17 cycles per day, transferring about 1000 gallons of manure per cycle. The propeller was added to the pump because solids were building up in the preheat tank. A sonar control on the two barn pumps and preheat tank allows flexibility in the manure volumes pumped.

The manure will continue to be analyzed monthly for the next year. The pH of the digester effluent was 7.0. The biogas has tested at 65% to 72% methane.

The anaerobic digestion system at Top Deck has been generating about 100 kW of electricity from manure from about 700 cows. It ran initially at about 95 kW and later peaked at 125 kW. Of the two generating options, the microturbines have had fewer mechanical problems than the engine/generator combination because there are fewer mechanical parts. It is good to have both systems because the microturbine can almost heat the digester and preheat tanks if the engine/generator system is down. The

30-kW microturbine runs about 25.5 kW when the weather is hot and increased to about 28.5 kW with cooler October weather.

Approximate costs for the digester and electricity generation system were as follows: Of the total $586,500 cost, the digester partners—Alliant Energy, the IDNR, and Top Deck—paid $335,000, $157,000, and $94,500 respectively.

There are several possible ways to benefit from the dairy manure digester. These benefits include the following:

- Generation of electricity
- Improved odor control on the manure storage
- Excess heated water from the engine/microturbine set
- Savings from the use of the digested manure solids for free stall barn bedding

It is hard to put a cost on improved odor control because the farm is not in a high-traffic area. The digested solids are not being separated and used yet. Therefore, the primary financial return is currently from electricity and heat.

Top Deck Holstein's average electric usage varies from 28,880 kilowatt-hour (kWh) in winter (cost = $1813) to its peak in the summer of 45,400 kWh ($3197). So the average savings for a month would be $2505 if they only generated electricity for their farm. This translates to a saving of $30,000 per year. The excess power generated would have been sold to the power company for 2 cents per kWh. Assuming that the electricity can be generated at 125 kW, the digester's best output so far, the amount of electricity potentially generated amounts to 3000 kWh per day or 90,000 kWh per month. If the farm uses an average of 37,100 kWh per month, the excess available for sale would be 52,900 kWh per month. At 2 cents per kWh, this would amount to an income of $1058 per month or $12,700 per year. The revenue would then be $30,000 (electricity used on farm) plus $12,700 (surplus electricity sold to power company) plus $4000 (saved on heat in the dairy center) for a total of $46,700 per year. Based on the $586,500, this would represent a simple return of 8.0%.

In the future, the manure solids separator has potential. This would essentially eliminate the farm's $30,000 annual bill for rice hull bedding. The separator and shed are estimated to cost $80,000. With the additional savings on bedding, the new annual return would be $76,700 on an investment of $666,500 ($586,500 + $80,000). The simple annual return would then be 11.5%.

Certainly, this economic exercise is an optimistic one. In reality, a portion of the revenue must go to depreciation, interest, repairs, taxes, insurance, and other cost factors. Also, the revenue and cost estimates do not recognize the effects of downtime and extraordinary repairs. Nevertheless, this simple analysis shows that anaerobic digestion has the potential to bring positive returns to dairy farms—from electricity saved, electricity sold, and recovery of heat and bedding.

CASE 11-3: EXTRACTS FROM "ECONOMIC AND FINANCIAL ASPECTS OF LANDFILL GAS TO ENERGY PROJECT DEVELOPMENT IN CALIFORNIA"*

Landfill Gas–to–Energy Overview

Landfill Gas Availability

LFG is produced by the anaerobic decomposition of organic waste in a landfill. Organic wastes include food waste, paper, wood, yard waste, and organic sludge. Municipal

---

* Source: SCS Engineers. 2002. *Economic and financial aspects of landfill gas to energy project development in California*. The California Energy Commission. Sacramento, CA.

solid waste contains a relatively large organic waste fraction. Industrial wastes, and therefore industrial landfills, generally contain much smaller fractions of organic waste. LFG collection, control, and utilization are, as a consequence, focused almost exclusively on municipal solid waste landfills.

LFG production begins shortly after waste is buried in a landfill and LFG will continue to be produced as long as organic waste is present. The decline in LFG production is gradual. In a dry climate, like Southern California, the rate of production will decline as little as 2% per year. In wetter climates, like Northern California, the rate of LFG production will decline at 6% per year.

Moisture is a significant factor in the rate of LFG production. The amount of moisture present in municipal solid waste does not vary appreciably in different regions in California, but additional moisture finds its way into the waste from precipitation. Landfills are designed to prevent the entry of water both during and after their active life; however, when the landfill is active, some water is inevitably added. The amount of water added is directly related to the precipitation in the region. LFG production can generally be correlated to the amount of annual precipitation in a region.

The most important factors affecting the amount of LFG produced from a fixed quantity of waste at any point in time are as follows:

- The quantity of waste (in tons)
- Its age (in years)
- The annual precipitation at the landfill (in inches)

There are several models which are available to project the amount of LFG that is being produced or will be produced in the future at a landfill. The most widely used model at the present time is a first-order model (sometimes called the Scholl Canyon Model). The U.S. Environmental Protection Agency's (EPA's) air emissions estimation model is a first-order model which is available at no cost. Copies of the model and operating instructions can be found at http://www.epa.gov/ttn/catc [Product Information, Software (executables and manuals), Landfill Gas Emissions Model (Version 2.01)].

While moisture is an important variable governing variations in LFG production, other factors play a role, including waste temperature, pH, and availability of nutrients. The waste management industry has recently focused research and development efforts on a landfilling concept known as a *bioreactor*. The bioreactor incorporates a series of cells of waste in which the principal parameters affecting waste decomposition are controlled with the intent of maintaining optimum conditions for waste degradation. The waste management industry sees several potential benefits from bioreactors, including quicker production of additional air space to support more waste disposal per acre and quicker stabilization of waste. The latter benefit would reduce long-term, post-closure maintenance costs of a landfill. The addition of liquid and its recirculation are common features of most bioreactor projects. The increased rate of waste degradation associated with bioreactors will increase the rate of LFG production. A bioreactor would allow a larger LFG-to-energy (LFGTE) project to be installed sooner; however, this benefit may be at the expense of LFG production in the future. In conventional landfills, it is assumed that the total amount of LFG which can be produced by a mass of waste is a fixed value. The fixed value is known as the *ultimate generation rate*, and is expressed as cubic feet per ton (ft^3/ton) or cubic meter per milligram. It is not known whether or not a bioreactor will increase total LFG production on a ft^3/ton basis. If it does not increase the ultimate generation rate, then the benefit of a bioreactor, from the perspective of LFGTE, is only to produce the fixed amount of LFG faster.

LFGTE Alternatives

LFG beneficial use can be grouped into three categories as follows:

- Medium-BTU gas production (sometimes called "direct use")
- Electric power generation
- Pipeline-quality gas production (sometimes called "high-BTU gas" production)

Electric power can be generated through the application of:

- Reciprocating engines
- Combustion turbines
- Steam cycle power plants
- Emerging technologies including microturbines, fuel cells, and Stirling engines
- Co-firing of LFG with fossil fuels in conventional electric power plants

Medium-BTU gas utilization is a concept through which the LFG is given minimal cleanup and is used to completely or partially displace a fossil fuel in boilers (commercial, institutional, and industrial), furnaces, and kilns. Co-firing of LFG with fossil fuel in conventional power plants is typically considered to be a medium-BTU LFG application, even though electric power is being produced.

High-BTU gas production involves extensive cleanup of the LFG to a level of quality so that it can be introduced into existing pipelines as a direct substitute for natural gas. High-BTU gas can also be compressed or liquefied and be used for vehicle fuel. Technologies currently in use for production of high-BTU gas include the membrane process, the solvent absorption process, and the molecular sieve process.

Figure 11-6 identifies the range of beneficial uses and technologies to be discussed in this section.

Medium-BTU Gas Utilization.   When LFG is used as a medium-BTU gas, it is directly used as a substitute for fossil fuel with very little treatment. The LFG is used at the methane content as seen at the landfill's flare station—which is about 40% to 55% methane. The LFG has an energy value of 400 to 550 BTU/ft^3 (higher heating value [HHV]). It can be blended with natural gas—which has an energy value of 1000 BTU/ft^3 (HHV) or it can be fired separately. The principal advantage associated with medium-BTU gas utilization is that the CO_2 does not need to be removed prior to LFG utilization. This results in a significant reduction in LFG-processing costs. The cost savings are partially offset by the need to construct a dedicated pipeline direct to the gas user and/or the need to modify the user's piping and fuel burning equipment to accommodate LFG firing. The cost of a dedicated pipeline can be largely eliminated if the potential LFG user is located at or adjacent to the landfill.

Medium-BTU gas has been successfully used at more than 50 locations in the United States. The applications include the following:

- Firing in commercial, institutional, and industrial boilers at colleges, hospitals, and several types of industries
- Firing in industrial furnaces, including cement kilns, aggregate dryers, ovens, and waste incinerators
- Firing in conventional electric power plants with coal or natural gas

The key to development of a successful medium-BTU gas project is identification of a fairly large, year-round user of fossil fuel which is not too distant from the landfill.

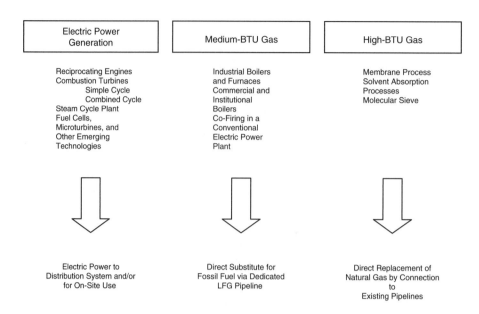

FIGURE 11-6 *Landfill gas (LFG) utilization alternatives. (Source: SCS Engineers. 2002.* Economic and financial aspects of landfill gas to energy project development in California. *The California Energy Commission. Sacramento, CA.)*

Figure 11-7 shows the standard process for production of medium-BTU gas. Compression is employed in order to: (1) reduce the diameter of the conveyance pipeline; (2) to overcome pressure losses as the gas moves through the conveyance pipeline; and (3) to supply an end point pressure suitable for the user's needs. Refrigeration is employed for advanced moisture removal to ensure that no condensate is formed in the conveyance pipeline and to produce a moisture-free gas for the end user.

If the end user's fuel specification is particularly demanding, then H_2S and/or non-methane organic compound (NMOC) removal can be added to the treatment process; however, the addition of such steps is unusual. Figures 11-8 and 11-9 illustrate these add-on processing steps.

Compression, cooling, and chilling result in increased production of LFG condensate and the generation of liquid hydrocarbon waste. The liquid hydrocarbon waste will consist of oil carried over from the compressors and hydrocarbons condensed from the LFG. The liquid hydrocarbon waste is usually not hazardous, but it must be sent to a proper disposal outlet.

Electric Power Generation
Reciprocating Engines.   Reciprocating engines are the most widely used prime movers for LFG-fired electric power generation. Waukesha, Superior, Caterpillar, and Jenbacher are the most commonly employed equipment suppliers. The capacity of the individual engines proven in LFG service varies from 0.1 to 3.0 MW. Reciprocating engines are manufactured in capacities much larger than 3 MW; however, the larger units have not been proven in LFG service. It is believed that the largest LFG-fired reciprocating engine-based power plant is in the United States, and has a net power output of 12 MW. There are more than 200 LFG-fired reciprocating engine power plants operating worldwide.

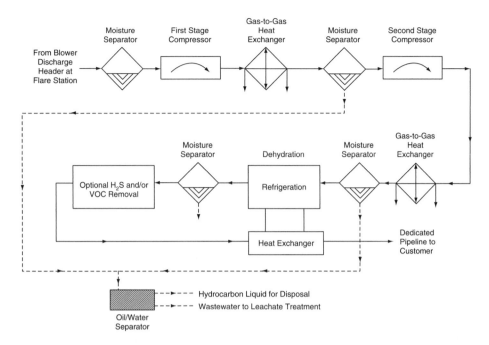

FIGURE 11-7   *Landfill gas cleanup process for medium-BTU gas. H_2S = hydrogen sulfide; VOC = volatile organic compound. (Source: SCS Engineers. 2002.* Economic and financial aspects of landfill gas to energy project development in California. *The California Energy Commission. Sacramento, CA.)*

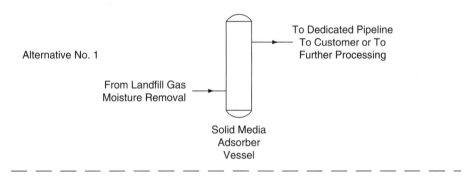

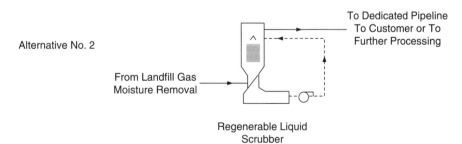

FIGURE 11-8   *Optional landfill gas cleanup processes for hydrogen sulfide (H_2S) removal. (Source: SCS Engineers. 2002.* Economic and financial aspects of landfill gas to energy project development in California. *The California Energy Commission. Sacramento, CA.)*

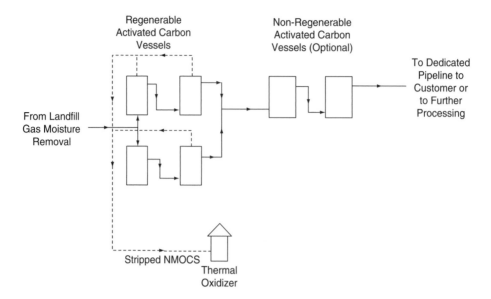

FIGURE 11-9   *Optional landfill gas cleanup process for non-methane organic compound (NMOC) removal. (Source: SCS Engineers. 2002.* Economic and financial aspects of landfill gas to energy project development in California. *The California Energy Commission. Sacramento, CA.)*

The principal advantage of reciprocating engines compared with other power-generation technologies is a better heat rate at lower capacities. An additional advantage of reciprocating engines is that the units are available in many different incremental capacities, which makes it easy to tailor the size of small plants to the specific rate of gas production at a landfill. Most small LFG power plants employ reciprocating engines.

An important disadvantage to reciprocating engines is that they produce higher emissions of nitric oxide (NO_x), carbon monoxide (CO), and NMOCs than other electric power–generation technologies. Significant progress has, however, been made in reducing NO_x emissions in recent years. A second disadvantage to reciprocating engines is that their operation/maintenance costs on a per-megawatt-hour basis are higher than for other power-generation technologies.

Station load for a reciprocating engine plant is about 7% of gross power output. The net heat rate for a typical reciprocating engine plant is 10,600 BTU/kWh (lower heating value [LHV]).

Typical air emissions for a reciprocating engine plant are shown in Table 11-2.

The jacket water coolers and lube oil coolers for reciprocating engines normally reject their heat through closed-loop, liquid-to-air heat exchangers. Wastewater is not produced in satisfying the plant's cooling requirements. Figure 11-10 is a schematic showing electric power generation with a reciprocating engine. In some cases, it may be possible to productively utilize the waste heat of a reciprocating engine plant.

Reciprocating engines generally require a relatively simple LFG-pretreatment process consisting of compression and removal of free moisture. Free moisture (water droplets) is removed by use of simple moisture separators (knockout drums), cooling of the LFG in ambient air-to-LFG heat exchangers, and coalescing-type filters. Moisture removal also removes particulates; however, LFG is generally fairly particulate free. Some of the NMOCs in the LFG are removed as a result of compression and cooling. Compression is usually provided by flooded screw-type blowers or centrifugal

TABLE 11-2   Typical Air Emissions for a Reciprocating Engine Plant

Emitted Agent	lbs/MMBTU
NO_x	0.200
CO	0.790
NMOCs	0.490
SO_x	0.008
Particulates	0.160

NO_x = nitric oxide; CO = carbon monoxide; NMOCs = non-methane organic compound; SO_x = sulfur oxide.
(Source: SCS Engineers. 2002. *Economic and financial aspects of landfill gas to energy project development in California.* The California Energy Commission. Sacramento, CA.)

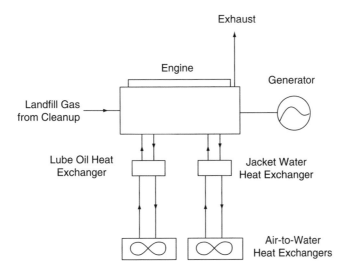

FIGURE 11-10   *Power generation with reciprocating engines. (Source: SCS Engineers. 2002. Economic and financial aspects of landfill gas to energy project development in California. The California Energy Commission. Sacramento, CA.)*

blowers. The reciprocating engines can require between 3 and 60 psig of fuel pressure. Figure 11-11 is a schematic showing the cleanup process for LFG for a reciprocating engine plant.

A few of the early LFG-fired reciprocating engine plants employed refrigeration units to chill the LFG to 40°F in order to induce additional moisture and NMOC condensation. It is also possible to use desiccant-type dryers instead of chillers and/or to employ activated carbon for advanced NMOC removal.

Engine manufacturers place restrictions on the amount of sulfur-bearing compounds and the total organic halide content which they will tolerate in the LFG. H_2S is the principal sulfur-bearing compound in LFG. Chlorine is present in some of the NMOCs found in LFG. Chlorinated compounds are responsible for virtually all of the organic halides in LFG. LFG infrequently exceeds the limits for H_2S and NMOCs imposed by reciprocating engine manufacturers; for this reason, a pretreatment scheme

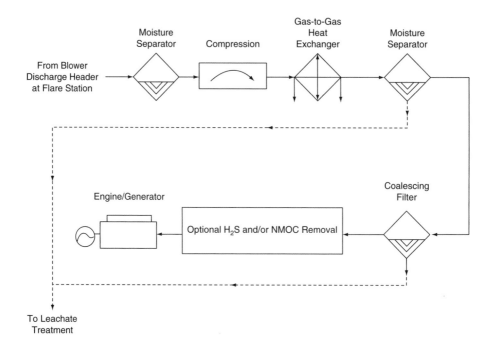

FIGURE 11-11 *Landfill gas cleanup process for reciprocating engines power generation.*
H_2S = *Hydrogen sulfide; NMOC = non-methane organic compound. (Source: SCS Engineers.*
2002. Economic and financial aspects of landfill gas to energy project development in California.
The California Energy Commission. Sacramento, CA.)

consisting of compression and simple moisture separation (knockout drum, air-to-LFG
heat exchanger, and coalescing filter) is virtually always the extent of LFG processing
at a reciprocating engine plant.

Combustion Turbines.   While less prevalent than reciprocating engines, combustion
turbines have seen widespread use as prime movers in LFG-fired electric power generation. The most widely used combustion turbine is the 3.3-MW Solar Centaur. The
Solar Saturn (1.0 MW) and Solar Taurus (5.2 MW) turbines have also been used in
LFG service. Virtually every LFG-fired combustion turbine installation is a simple-
cycle installation.
    The principal advantages of the combustion turbine as compared to a reciprocating
engine are its lower air emissions and lower operation/maintenance costs. The principal drawback to the combustion turbine is its high net heat rate. The poor net heat
rate owes itself to two factors. First, the station power for a combustion turbine–based
plant is about 15% of gross power output as compared to about 7% for a reciprocating
engine–based plant. The combustion turbines require a much higher gas pressure, which
increases the power consumption of the fuel gas compressors. Second, the combustion
turbines used in LFG electric power production are small and are not as efficient as the
larger units commonly employed in the electric power industry. The largest LFG-fired
combustion turbine plant is believed to be in the United States and consists of five Solar
Centaurs with a gross capacity of 16.5 MW. Solar has more experience with LFG than
any other combustion turbine manufacturer. There are more than 35 combustion turbines operating on LFG at more than 20 power plants.
    Station load for a simple cycle combustion turbine plant is about 15%, and net
heat rates vary from 12,200 to 16,400 BTU/kWh (LHV). The larger, new combustion
turbines are more fuel efficient.

Combustion turbines have traditionally achieved low NO_x emission rates based on water injection, steam injection, selective catalytic reduction (SCR), or dry low-NO_x burner technology. None of these technologies have been applied to LFG due to technical/operational concerns, and due to the fact that NO_x emissions when firing on LFG are lower than when firing on natural gas under otherwise identical conditions. Expected air emission rates for NO_x, CO, and NMOC, when firing on LFG in a Solar combustion turbine, are shown in Table 11-3.

The combustion turbine/generator and the fuel gas compressors normally reject their heat through closed-loop, liquid-to-air heat exchangers. Wastewater is not produced during cooling.

Virtually all combustion turbine installations to date have been simple-cycle installations. Simple-cycle plants have been preferred because the power plants have been relatively small and because LFG is an inexpensive fuel. Figure 11-12 contains schematics showing simple-cycle and combined-cycle combustion turbine configurations.

Combustion turbines require a higher-pressure fuel supply than reciprocating engines. The required fuel supply pressure is in the range of 150 to 250 psig. Two stages of LFG compression are employed. Particulate in the LFG has sometimes caused problems with the combustion turbine's fuel injection nozzles. A small water wash scrubber is normally provided in the pretreatment process to prevent this problem. Figure 11-13 is a schematic showing the cleanup process for LFG for a combustion turbine-based power plant.

If required by a combustion turbine manufacturer, H_2S and/or NMOCs can be removed. Solar has not required H_2S or NMOC removal in installations to date. Removal of these compounds may be required if a less experienced combustion turbine manufacturer is employed or if environmental regulations require installation of an SCR for NO_x or CO control. Activated carbon is normally employed to remove compounds which would otherwise cause SCR catalyst fouling. H_2S can be removed in a solid media absorber vessel (containing an iron sponge or a proprietary compound such as Sulfatreat) or in a liquid scrubber. Figures 11-8 and 11-9 are schematics showing H_2S and VOC removal processes.

Steam Cycle Power Plants.   Conventional boilers with steam turbines have seen limited application in LFG-fired electric power production. It is believed that eight steam-cycle power plants are operating on LFG worldwide. Most LFG-fired power plants are less than 10 MW in capacity, which puts the steam cycle at a cost disadvantage when compared against reciprocating engines and combustion turbines. The steam-cycle power

TABLE 11-3   Expected Air Emission Rates When Firing on Landfill Gas in a Solar Combustion Turbine

Emitted Agent	*lbs/MMBTU*
NO_x	0.120
CO	0.090
NMOCs	0.015
SO_x	0.008
Particulates	0.160

NO_x = nitric oxide; CO = carbon monoxide; NMOCs = non-methane organic compound; SO_x = sulfur oxide.
(Source: SCS Engineers. 2002. *Economic and financial aspects of landfill gas to energy project development in California*. The California Energy Commission. Sacramento, CA.)

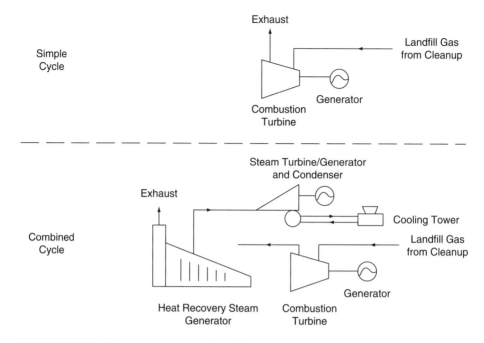

FIGURE 11-12   *Power generation with combustion turbines. (Source: SCS Engineers. 2002.* Economic and financial aspects of landfill gas to energy project development in California. *The California Energy Commission. Sacramento, CA.)*

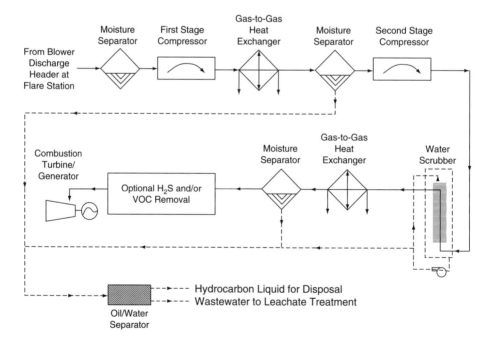

FIGURE 11-13   *Landfill gas cleanup process for combustion turbine power generation.* H_2S = *Hydrogen sulfide. (Source: SCS Engineers. 2002.* Economic and financial aspects of landfill gas to energy project development in California. *The California Energy Commission. Sacramento, CA.)*

plant becomes more cost competitive as the size of the plant increases. The steam-cycle power plant offers lower air emissions than either reciprocating engines or combustion turbines. As a consequence, steam cycles have been given preferential treatment in regions with stringent air quality regulations, even when the size of the plant was relatively small.

The smallest operating steam-cycle power plant is a 6-MW plant at the BKK Landfill (West Covina, CA) and the largest is a 50-MW plant at the Puente Hills Landfill (Whittier, CA). The 50-MW power plant at the Puente Hills Landfill has been in operation for almost 15 years and has been extremely reliable, demonstrating a capacity factor of over 96%.

The heat rate of a steam-cycle power plant is dependent on the details of the power cycle as established by the design engineer. The most efficient units operate at a gross heat rate of about 10,100 BTU/kWh (HHV). The least efficient units operate at gross heat rates as high as 15,200 BTU/kWh (HHV). Station load is in the neighborhood of 8%. Net heat rates are, therefore, in the range of 11,000 to 16,500 BTU/kWh (HHV). The most efficient steam cycles use higher temperature and pressure (1000°F/1350 psig), air preheaters, and up to five stages of feed-water heating. The least efficient units operate at low temperature and pressure (650°F/750 psig) and are not equipped with air preheaters or feed-water heaters.

NO_x emissions, when firing on LFG, are roughly half the NO_x emissions when firing on natural gas, with other conditions the same. Low levels of NO_x can be achieved through the application of recycle flue gas. Air emissions for a steam-cycle power plant employing recycle flue gas are shown in Table 11-4.

The wastewater generated by an LFG-fired steam-cycle power plant is identical to that of a natural gas–fired steam-cycle power plant. The wastewater includes boiler blowdown, wastewater from boiler makeup water treatment, and cooling tower blowdown. Figure 11-14 is a schematic showing the steam-cycle power plant concept.

LFG requires no pretreatment prior to firing in a conventional boiler. LFG is normally taken from the discharge side of the LFG blowers in the landfill's flare station. Large water droplets and particulates have already been removed in the flare station's moisture separator. LFG pressure is increased to the pressure required by the boiler's burners by a set of LFG booster blowers to the pressure range of 1 to 4 psig.

Fuel Cells.   The general public was first introduced to fuel cells in the 1960s when fuel cells began to provide internal power for manned spacecraft. Fuel cells chemically convert hydrogen and oxygen to electricity while emitting water vapor and CO_2. Tanks of hydrogen and oxygen supply the feedstock for spacecraft applications. In terrestrial applications, the oxygen is supplied by the ambient air and hydrogen is produced from methane or other hydrogen-containing feedstock. Fuel cells have interested the power-generation industry and regulators due to their high fuel efficiency and ultra low emissions.

There are several types of fuel cells either available or under development including phosphoric acid type, molten carbonate type, solid oxide type, and polymer-membrane type. The phosphoric acid type is commercially available. An International Fuel Cells Corporation subsidiary, the ONSI Corporation, has shipped more than 200 of their PC25 package fuel cells since their introduction in 1991.

The 200-kW PC25 package includes three steps in a 10-foot-wide by 20-foot-long by 10-foot-high box:

- A fuel processor in which natural gas is converted to a hydrogen-rich gas using steam reformer technology
- A power section in which hydrogen is combined with oxygen (from the air) to produce direct current power, water, CO_2, and heat

TABLE 11-4   Air Emissions for a Steam-Cycle Power Plant
Employing Recycle Flue Gas

Emitted Agent	lbs/MMBTU
NO_x	0.03
CO	0.01
NMOCs	0.005
SO_x	0.008
Particulates	0.02

NO_x = nitric oxide; CO = carbon monoxide; NMOCs = non-methane organic compound; SO_x = sulfur oxide.
(Source: SCS Engineers. 2002. *Economic and financial aspects
of landfill gas to energy project development in California.* The
California Energy Commission. Sacramento, CA.)

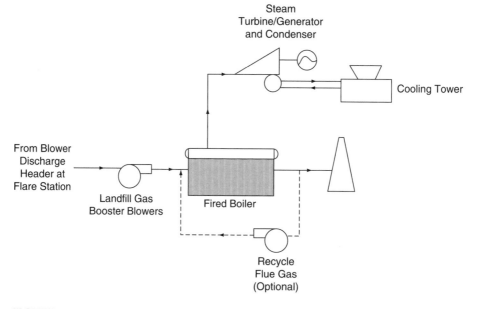

FIGURE 11-14   *Power generation with steam cycle. (Source: SCS Engineers. 2002.* Economic
and financial aspects of landfill gas to energy project development in California. *The California
Energy Commission. Sacramento, CA.)*

- A power conditioner where direct current power is converted to alternating current
  power

If it is possible to put the heat to a productive use, then total efficiency of the fuel
cell can be further enhanced.

The fuel cell is considered an opportunity for LFG utilization since it contains methane, the feedstock for stationary fuel cell applications. There have been two relatively
short-term but successful fuel cell demonstration tests to date. There is one commercially operating unit. Fuel cells are nevertheless attractive to the LFG utilization industry
because: (1) they are available in small incremental capacities (making them applicable
to projects smaller than possible with other power-generation technologies); (2) they
produce almost zero emissions of criteria pollutants and produce little noise; (3) they
can operate with little supervision; and (4) they are believed to have moderately low

operating costs. The principal obstacle to widespread application for projects in the 200 kW to 2 MW range is high capital cost.

LFG cleanup is an important issue. Commercially available fuel cell packages employ catalysts that would be fouled by trace compounds in LFG.

A LFG cleanup system for a fuel cell would include the following:

- An adsorber for H_2S removal
- Chilling and desiccation (to remove moisture and some hydrocarbons)
- Activated carbon to adsorb remaining trace organics

Microturbines. The microturbine is a recently commercialized distributed generation technology. As of June 2001, two companies manufacture and sell microturbines—Capstone Turbine Company (Chatsworth, CA) and Honeywell Power Systems (Albuquerque, NM). Capstone currently offers a 30-kW and a 60-kW unit. Honeywell currently offers a 75-kW unit. Capstone has delivered more than 3000 units. Honeywell has delivered more than 300 units. At least three other microturbine manufacturers will soon offer units for general sale—Elliot Energy Systems (Secure Power); Ingersoll Rand (NREC Energy Systems); and ABB/Volvo. It is expected that these manufacturers will release units in the 45- to 100-kW range within the next 1 to 2 years.

Most microturbine installations to date have employed natural gas as their fuel. Permanent (versus experimental) microturbine installations have also burned oil field flare gas, municipal wastewater-treatment plant digester gas, and LFG. As of June 2001, there are about 50 microturbines operating on these waste fuels. An additional 100 units were being installed and are expected to be operational as of September 2001. As of June 2001, the longest run time for a microturbine on natural gas was about 16,000 hours. The longest microturbine run time on waste fuel was about 8000 hours. The longest run time on LFG was about 2000 hours.

The microturbine is a derivative of the much larger combustion turbines employed in the electric power and aviation industries. Combustion air and fuel are mixed in a combustor section, and the release of heat causes the expansion of the gas. The hot gas is sent through a gas turbine, which is connected to a generator. The units are normally equipped with a recuperator, which heats the combustion air using turbine exhaust gas in order to increase the unit's overall efficiency. The combustion air is compressed using a compressor driven by the gas turbine. The fuel must be supplied to the combustor at 70 to 80 psig. In some natural gas–fired applications, the gas is available at this pressure from the pipeline. In LFG applications, a gas compressor is required. The microturbine differs from traditional combustion turbines in that the microturbine spins at a much faster speed. The microturbines that are now on the market are equipped with air bearings rather than traditional mechanical bearings in order to reduce wear.

A typical LFG-fired microturbine installation would have the following components:

- LFG compressor(s)
- LFG pretreatment equipment
- Microturbine(s)
- Motor control center
- Switchgear
- Step-up transformer

Microturbines require about 13,900 BTU/kWh (LHV) of fuel on a gross power output basis. Station load is about 15%, resulting in a net power output of about 16,350 BTU/kWh (LHV).

Microturbines are most applicable where the following circumstances exist:

- Low quantities of LFG are available.
- The LFG has a low methane content.
- Air emissions are of great concern.
- Emphasis is being placed on satisfaction of on-site power requirements, rather than exporting power.
- A requirement for hot water exists at or near the landfill.

Microturbines can operate on LFG with a methane content of 35% (and perhaps as low as 30%). A 75-kW unit requires less than 50 standard cubic feet per minute (scfm) of LFG (at 35% methane content). Microturbines can be used at small landfills and at old landfills where LFG quality and quantity would not support more traditional LFG electric power-generation technologies.

Air emissions from a microturbine are much lower than for a reciprocating engine. Microturbines have demonstrated NO_x emissions less than one tenth those of the best performing reciprocating engines. The NO_x emissions from microturbines are lower than the NO_x emissions from an LFG flare. NO_x emissions of less than 0.01 lbs/MMBTU have been demonstrated by microturbines fired on LFG.

It is possible to produce hot water (up to 200°F) from the waste heat in the microturbine exhaust. Microturbine manufacturers offer a hot water generator as a standard option. Landfills in colder climates probably have a space-heating requirement—which is often satisfied by a relatively expensive fuel (such as propane). Hot water users (such as hotels, industrial or institutional buildings, etc.) are sometimes adjacent to landfills—particularly closed landfills. The sale or use of microturbine waste heat can significantly enhance project economics.

Microturbines have the following advantages as compared to reciprocating engines:

- Lower air emissions
- Availability in smaller incremental capacities
- Ability to burn a lower methane content LFG

Disadvantages of microturbines as compared to reciprocating engines include the following:

- A higher heat rate (about 35% more fuel consumed per kWh produced)
- Limited experience on LFG

The higher heat rate of the microturbine is generally not an issue since LFG is waste fuel.

Pipeline-Quality Gas Production.   LFG typically contains about 40% to 55% methane when it reaches the landfill's flare station, with the balance of the gas consisting primarily of CO_2 and secondarily of air (nitrogen plus oxygen) plus water vapor. LFG also contains trace compounds including NMOCs (such as toluene, trichloroethylene, and vinyl chloride) and H_2S. LFG has an HHV of about 400 to 550 BTU/ft³. LFG can be used to displace natural gas use in two ways. First, it can be subjected to light cleanup and be transmitted to an end user through a dedicated pipeline. The product gas retains its original energy content and the LFG displaces or is blended with natural gas at its point of use. Natural gas has a heating value of about 1000 BTU/ft³ (HHV). As discussed above, this "direct use" of LFG is commonly known as "medium-BTU" gas use. A second way to displace natural gas is to inject it into an existing natural gas distribution network. Natural gas, as distributed through pipelines to customers, must meet strict quality standards. Pipeline operators will allow LFG to enter their pipelines only after the LFG has been processed to increase its energy content and to meet strict standards for H_2S, moisture, CO_2, and NMOCs. The need to roughly double the energy content of LFG has led the LFG utilization industry to call gas beneficiated to pipeline quality "high-BTU" gas.

A typical pipeline-quality gas specification is as follows:

- Heating value (HHV) > 970 BTU/ft^3
- H_2S < 4 parts per million by volume (ppmv)
- Water vapor < 7 lbs/million ft^3
- Oxygen < 0.4%
- CO_2 < 3%
- Nitrogen plus CO_2 < 5%

The 970 BTU/ft^3 (HHV) limitation requires, in effect, that oxygen plus CO_2 plus nitrogen be limited to less than 3%. The product gas must also be free of environmentally unacceptable substances and must be pressurized to the pressure of the pipeline to which the gas production facility is interconnected. Pipeline pressure typically varies from 100 to 500 psig.

The following steps must be taken to convert LFG to pipeline-quality gas:

- Prevention of air infiltration into the LFG well field
- Moisture removal
- Sulfur removal
- NMOC removal
- CO_2 removal

The removal of CO_2 is the principal step taken to increase energy content. The prevention of air infiltration into the well field is also a critical step, not only because air infiltration reduces energy content, but also because it is necessary to satisfy tight product gas nitrogen and oxygen limitations. The addition of processing steps to remove nitrogen and oxygen from the LFG is widely viewed as being prohibitively expensive. At most landfills, elimination of air infiltration will require that the utilization facility be supplied by wells located in the "core" of the landfill. A separate perimeter LFG-collection system must often be operated for gas migration control. Each well on the core gas system must be carefully monitored to maintain as close as practical "zero" air infiltration operation.

CO_2 can be removed from LFG using three well-proven technologies: the membrane process, the molecular sieve, and solvent absorption.

Membrane Process.   The membrane process exploits the fact that gases, under the same conditions, will pass through polymeric membranes at differing rates. A "fast" or highly permeable gas such as CO_2 will pass through a membrane approximately 20 times faster than a "slow" or less permeable gas such as methane. Pressure is the driving force for the separation process. The feed gas (LFG) and product gas (predominately methane) enter and exit the membrane module at approximately the same pressure. The permeate gas (predominately CO_2) exits at a lower pressure. The operating pressures, number of membrane stages in series, and provisions for gas recycle depend on desired methane recovery percentage and product gas methane purity. In natural gas–processing applications, both methane recovery percentage and desired product gas methane purity are highly important. In LFG applications, product gas methane purity is of paramount importance; however, methane recovery as a percentage of methane in the raw LFG is of less importance. The membrane configuration employed for an LFG-utilization project should strike the optimal balance between capital cost and methane recovery on a given project. Total methane recovery is normally about 85% and the product gas contains 97% methane.

Figure 11-15 is a diagram of the membrane-separation process. Two stages of membranes plus recycle of second-stage permeate generally represents an optimal configuration for a pipeline-quality gas project.

The membranes must be protected against moisture, NMOCs, and particulates. These impurities can harm the membranes and reduce their life. Figure 11-16 shows the process which has been used to cleanup LFG prior to processing the LFG through the membrane process. It is necessary to compress the LFG to about 600 psig. This is accomplished through two stages of compression. Moisture is first removed through moisture separators and post-compression cooling of the LFG. The compressors can be driven by electric motors or by reciprocating engines fired on raw LFG, pretreated LFG, or product gas. Refrigeration is used to achieve advanced moisture removal. A separate H_2S removal step is added only if the LFG hydrogen concentration exceeds 60 ppmv. If the H_2S concentration is less than 60 ppmv, then the membrane alone can meet the 4-ppmv H_2S product gas specification. The activated carbon step removes NMOCs not condensed by the time the LFG reaches the activated carbon vessels. The activated carbon is regenerated on-site and the NMOC-laden waste stream is directed to a thermal oxidizer for destruction of the NMOCs. The reject CO_2 stream from the membranes is also directed to the same thermal oxidizer where auxiliary fuel (typically LFG) may or may not be required to support high-temperature combustion.

Molecular Sieve.   The molecular sieve is a vessel which contains a media which preferentially adsorbs certain molecules (in this case, CO_2) when contacted with a gas stream which is under pressure. When the adsorption capacity of the media is exhausted, the vessel is brought off-line and is regenerated through a depressurization and purge cycle. The CO_2 exhaust stream from the on-line molecular sieve vessels is used for purge. The stream is backflowed through the off-line molecular sieve to carry the waste stream to a thermal oxidizer. In some instances, the waste stream can be discharged to the atmosphere. The thermal oxidizer generally requires some supplemental energy, which can be provided by LFG or product gas.

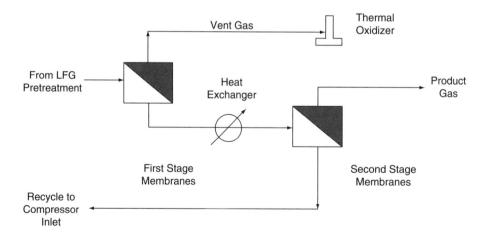

FIGURE 11-15 *High-BTU gas by membrane process. LFG = landfill gas. (Source: SCS Engineers. 2002.* Economic and financial aspects of landfill gas to energy project development in California. *The California Energy Commission. Sacramento, CA.)*

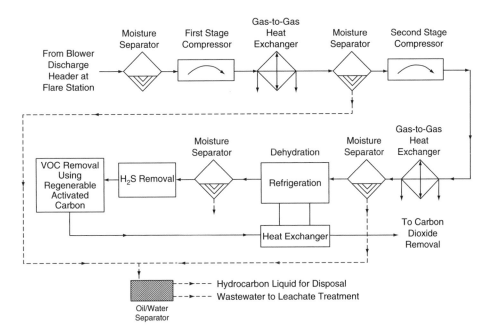

FIGURE 11-16   *Landfill gas cleanup process for high-BTU gas. H_2S = hydrogen sulfide. (Source: SCS Engineers. 2002.* Economic and financial aspects of landfill gas to energy project development in California. *The California Energy Commission. Sacramento, CA.)*

The molecular sieve relies on adsorption. Adsorption is a phenomenon whereby molecules in a fluid phase spontaneously concentrate on a solid surface without undergoing any chemical change. Adsorption takes place because forces on the surface of the adsorption media attract and hold the molecules that are to be removed. In the LFG-processing application, a medium is employed which prefers CO_2. The adsorbed CO_2 is released from the surface of the medium when the medium is depressurized. The molecular sieve process is also known as pressure swing adsorption.

Figure 11-17 is a schematic diagram showing the molecular sieve process.

The adsorption media employed in the molecular sieve process must be protected against contaminants in the LFG. The LFG-pretreatment process for the molecular sieve is virtually identical to the pretreatment process employed for the membrane process. A separate H_2S removal step is required, however, since the molecule sieve is optimized for CO_2 removal.

Absorption Processes.   A typical absorption process plant employs a liquid solvent to scrub NMOCs and CO_2 from the LFG. A solvent called Selexol is most frequently used. A typical absorption process plant employs the following steps:

• LFG compression (using electric drive, LFG-fired engine drive, or product gas–fired engine drive)
• Moisture removal (using refrigeration)
• H_2S removal in a solid media bed (using an iron sponge or a proprietary media such as Sulfatreat) or by liquid scrubbing
• NMOC removal in a primary Selexol absorber
• CO_2 removal in a secondary Selexol absorber

In the Selexol absorber tower, the LFG is placed in intimate contact with the Selexol liquid. Selexol is a physical solvent which preferentially absorbs gases into the

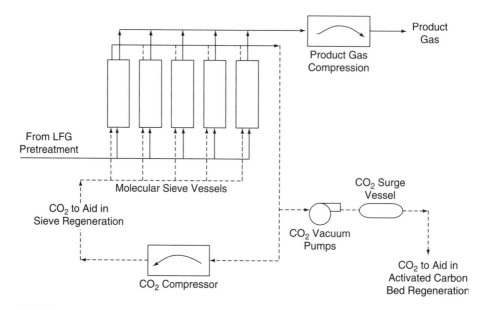

FIGURE 11-17   *High-BTU gas by molecular sieve. LFG = landfill gas; CO_2 = carbon dioxide. (Source: SCS Engineers. 2002. Economic and financial aspects of landfill gas to energy project development in California. The California Energy Commission. Sacramento, CA.)*

liquid phase. NMOCs are generally hundreds to thousands of times more soluble than methane. CO_2 is about 15 more times soluble than methane. Solubility is also enhanced with pressure. The above principles are exploited to remove VOCs and CO_2 from the LFG to yield a purified methane stream. The Selexol vessels operate at pressures in the range of 500 psi. The Selexol liquid is regenerated by lowering its pressure (flashing) and then running air through the depressurized Selexol to strip off the VOCs and CO_2. The stripper air is normally sent to a thermal oxidizer where all or part of the thermal energy required to support combustion is supplied by the VOCs and methane in the stripper air.

Overview of LFGTE Technologies in California

Existing Projects.   Table 11-5 lists LFGTE projects operating in California as of July 2001. There were 38 electric power-generation projects totaling a little over 200 MW. There were six medium-BTU projects operating as of July 2001.

The electric power projects include five steam-cycle power plants, five simple-cycle combustion turbine plants, and 28 reciprocating engine plants. The largest power plant was a 50-MW steam-cycle power plant at Puente Hills Landfill in Whittier.

The largest medium-BTU project is at the Scholl Canyon Landfill. The LFG is compressed, refrigerated, desulfurized, and is sent through an 8-mile pipeline to the City of Glendale's conventional steam-cycle power plant, where it is co-fired with natural gas to produce about 10 MW.

Planned and Potential Projects.   Table 11-6 lists planned and potential LFGTE projects in California. Planned and potential projects could double the amount of electric power now being produced by LFG in California.

TABLE 11-5   Existing Landfill Gas (LFG)-to-Energy Projects in California

Landfill Name	City	Current Status	MSW in Place (tons)	Landfill Owner Type	Power Generation Capacity		Medium-BTUt Capacity	
					MW	MWh/ Yr	MM scfd	MM BTU/Yr
Acme Landfill	Martinez	Operational	10,700,000	Private		—	1.8	328,500
Altamont Landfill	Livermore	Operational	26,450,512	Private	6.6	57,816.00		—
American Canyon Landfill	Napa	Operational	4,200,000	Public	1.5	13,140.00		—
Austin Road Landfill	Stockton	Operational	2,030,690	Public	0.8	7,008.00		—
Bailard Landfill	Oxnard	Operational	4,336,609	Public	0.0	—		—
BKK Landfill-Phase I	West Covina	Operational	46,000,000	Private	6.0	52,560.00		—
BKK Landfill-Phase II	West Covina	Operational	46,000,000	Private	6.0	52,560.00		—
Bradley Landfill	Sun Valley	Operational	23,000,000	Private	0.0	—		—
Central Landfill	Petaluma	Operational	9,575,690	Public	3.2	28,032.00		—
City of Santa Clara Landfill	Santa Clara	Operational	3,500,000	Public	1.4	12,264.00		—
City of Sunnyvale	Sunnyvale	Operational	2,300,000	Public	1.6	14,016.00		—
Coastal Landfill	Oxnard	Operational	3,300,000	Public	0.0	—		—
Coyote Canyon Landfill	Irvine	Operational	25,000,000	Public	17.0	148,920.00		—
Crazy Horse Landfill	Prunedale	Operational	1,855,351	Public	1.4	12,264.00		—
Davis Street Landfill	San Leandro	Operational	9,700,000	Public	—		1.4	255,500
Guadalupe Disposal Site	San Jose	Operational	5,500,000	Private	2.5	21,900.00		—

(Continued)

TABLE 11-5   Existing Landfill Gas (LFG)-to-Energy Projects in California—Cont'd

Landfill Name	City	Current Status	MSW in Place (tons)	Landfill Owner Type	Power Generation Capacity		Medium-BTUt Capacity	
					MW	MWh/ Yr	MM scfd	MM BTU/Yr
Industry Hills	Industry	Operational	1,500,000	Public	—		0.5	91,250
Lopez Canyon Landfill	San Fernando	Operational	16,700,000	Public	6.0	52,560.00	—	
Menlo Park Landfill	Menlo Park	Operational	5,000,000	Public	2.0	17,520.00	—	
Miramar Landfill– Phase I	San Diego	Operational	29,000,000	Public	6.0	52,560.00	—	
Miramar Landfill– Phase II	Sand Diego	Operational	29,000,000	Public	4.0	35,040.00	—	
Monterey Peninsula Landfill	Marina	Operational	5,500,000	Public	1.2	10,512.00	—	
Mountain gate Landfill	Los Angeles	Operational	10,000,000	Private		—	3.0	547,500
Newby Island Landfill– Phase I	San Jose	Operational	19,900,000	Private	5.0	43,800.00	—	
Olinda Landfill– Phase I	Brea	Operational	32,981,156	Public	5.7	49,932.00	—	
Otay Landfill	Chula Vista	Operational	8,099,138	Private	3.7	32,412.00	—	
Palo Alto Landfill	Palo Alto	Operational	2,700,000	Public	2.0	17,520.00	—	
Palos Verdes Landfill	Rolling Hills	Operational	23,000,000	Public	9.0	78,840.00	—	
Penrose Landfill	Sun Valley	Operational	9,000,000	Private	9.3	81,468.00	—	
Placer County Western Regional	Roseville	Operational	2,300,000	Public	1.1	9,636.00	—	
Prima Deshecha Landfill	San Juan Capistrano	Operational	7,400,000	Public	6.0	52,560.00	—	

(Continued)

TABLE 11-5   Existing Landfill Gas (LFG)-to-Energy Projects in California—Cont'd

Landfill Name	City	Current Status	MSW in Place (tons)	Landfill Owner Type	Power Generation Capacity		Medium-BTUt Capacity	
					MW	MWh/ Yr	MM scfd	MM BTU/Yr
Puente Hills Landfill (Steam Cycle)	Whittier	Operational	78,142,152	Public	50.0	438,000.00		—
Puente Hills (Gas Turbines)	Whittier	Operational	78,142,152	Public	4.0	35,040.00		—
Puente Hills (Rio Hondo)	Whittier	Operational	78,142,152	Public		—	0.2	36,500
Sacramento City Landfill	Sacramento	Operational	3,900,000	Public		—	1.2	219,000
San Marcos Landfill–Phase I	San Marcos	Operational	11,494,445	Public	1.3	11,388.00		—
Santa Clara Landfill	Oxnard	Operational	270,000	Public	5.6	49,056.00		—
Santa Cruz City Landfill	Santa Cruz	Operational	2,040,427	Public	0.7	6,132.00		—
Scholl Canyon Landfill	Glendale	Operational	22,511,265	Public		—	9.0	1,000,000
Sheldon-Arleta Landfill	Sun Valley	Operational	5,500,000	Public	0.0	—		—
Spadra Landfill	Pomona	Operational	12,500,000	Private	9.6	84,096.00		—
Sycamore Landfill	Santee	Operational	9,163,739	Public	1.5	13,140.00		—
Tajiguas Landfill	Santa Barbara	Operational	7,000,000	Public	3.8	33,288.00		—
Temescal Road Landfill	Corona	Operational	4,000,000	Public	2.0	17,520.00		—
Toyon Canyon Landfill	Los Angeles	Operational	16,000,000	Public	9.0	78,840.00		—

(Continued)

TABLE 11-5   Existing Landfill Gas (LFG)-to-Energy Projects in California—Cont'd

Landfill Name	City	Current Status	MSW in Place (tons)	Landfill Owner Type	Power Generation Capacity		Medium-BTUt Capacity	
					MW	MWh/ Yr	MM scfd	MM BTU/Yr
Visalia Disposal Site	Visalia	Operational	3,500,000	Public	1.6	14,016.00		—
West Contra Costa Landfill	Contra Costa	Operational	10,105,056	Private	3.0	26,280.00		—
Woodville Disposal Site	Woodville	Operational	1,580,000	Public	0.8	7,008.00		—
Yolo County Central Landfill	Davis	Operational	8,427,340	Public	2.4	21,024.00		—
					204.3	1,789,668.00	17.1	2,478,250

MSW = municipal solid waste; MMscfd = million standard cubic feet per day.
(Source: SCS Engineers. 2002. *Economic and financial aspects of landfill gas to energy project development in California.* The California Energy Commission. Sacramento, CA.)
Notes:
(1) Bradley Landfill and Sheldon-Arleta Landfill send LFG to a power plant at Penrose Landfill.
(2) Bailard Landfill and Coastal Landfill send LFG to a power plant at Santa Clara Landfill.
(3) Scholl Canyon's LFG is conveyed to the City of Glendale's Grayson Station, where it is co-fired with natural gas to produce about 10 MW in a 50-MW steam cycle plant.
(4) Mountaingate's LFG is conveyed to UCLA, where it is co-fired with natural gas to produce about 6 MW in a 16-MW combustion turbine cogeneration facility.

Benefits of LFGTE

The benefits of LFG collection and control/utilization are as follows:

- A reduction in GHG emissions by eliminating the uncontrolled release of methane to the atmosphere
- A reduction in the potential for explosions in structures at or near a landfill
- A reduction in odor emissions
- A reduction in emissions of hazardous organic air pollutants to the atmosphere
- The recovery of a low-cost, relatively high-quality fuel

The use of recovered LFG as a fuel results in additional benefits, including the following:

- Reduced consumption of fossil fuel, a finite resource, with a renewable resource
- A further reduction in GHG emissions through deferral of consumption of fossil fuel
- A reduction in energy cost to the user and/or reduction in net operating costs to the landfill owner (depending on the size and type of the LFGTE project)

Public Health and Safety Issues. Not all LFG is immediately emitted to the atmosphere. A portion can travel underground and accumulate in basements and other

TABLE 11-6 Potential Landfill Gas (LFG)-to-Energy Projects in California

Landfill Name	City	Current Status	MSW in Place (tons)	Landfill Owner Type	Power Generation Capacity		Medium-BTU Capacity	
					MW	MWh /Yr	MM scfd	MMBTU /Yr
Amador Country Landfill	Ione	Potential	1,013,553	Public	0.8	7008		—
American Canyon– Phase II	Napa	Potential	482,541	Public	0.8	7008		—
Anderson	Anderson	Potential	1,554,000	Private	1.6	14,016		—
Arvin Landfill	Arvin	Potential	3,700,000	Public	2.4	21,024		—
Azusa Land Reclamation	Azusa	Planned	7,900,000	Private	3.0	26,280		—
B & J Landfill	Vacaville	Potential	780,000	Private	1.6	14,016		—
Badlands Disposal Site	Moreno Valley	Planned	1,720,000	Public	0.8	7008		—
Bakersfield Metro- politan	Bakers- field	Potential	1,405,607	Public	1.2	10,512		—
Ben Lomond Solid Waste	Ben Lomond	Potential	1,200,000	Public	0.8	7008		—
Blythe Disposal Site	Blythe	Planned	3,864,000	Public	1.6	14,016		—
Bonzi Landfill	Modesto	Potential	1,800,000	Private	0.8	7008		—
Bradley Landfill	Los Angeles	Potential	23,000,000	Private	6.0	52,560		
Buena Vista Disposal Site	Watson- ville	Planned	1,250,046	Public	0.8	7008		—
Calabasas Landfill	Agoura Hills	Planned	10,000,000	Public	10.0	87,600		—
California Street Landfill	Redlands	Potential	960,000	Public	0.8	7008		—
Chateau Fresno Landfill	Fresno	Potential	13,451,139	Private		—	3	547,500

(Continued)

TABLE 11-6   Potential Landfill Gas (LFG)-to-Energy Projects in California—Cont'd

Landfill Name	City	Current Status	MSW in Place (tons)	Landfill Owner Type	Power Generation Capacity		Medium-BTU Capacity	
					MW	MWh /Yr	MM scfd	MMBTU /Yr
Chestnut Avenue Landfill	Fresno	Potential		Private	0.8	7008		—
China Grade Landfill	Bakersfield	Potential	1,639,000	Public	0.8	7008		—
Chiquita Canyon Landfill	Valencia	Planned	10,500,000	Private	7.6	66,576		—
City Garbage (Cummings)	Eureka	Potential	1,740,000	Private	2.5	21,900		—
City of Paso Robles Landfill	Paso Robles	Potential	870,000	Public	0.8	7008		—
City of Santa Maria	Santa Maria	Potential	1,809,995	Public	0.8	7008		—
City of Ukiah	Ukiah	Potential	2,200,000	Public	1.6	14,016		—
Coachella Valley Disposal Site	Coachella	Planned	3,237,000	Public	1.6	14,016		—
Cold Canyon Landfill	San Luis Obispo	Potential	2,254,575	Private	1.6	14,016		—
Colton Refuse Disposal Site	Colton	Operational	6,500,000	Public	3.0	26,280		—
Corral Hollow Landfill	Tracy	Potential	690,000	Public	0.8	7008		—
Crescent City Landfill	Crescent City	Potential	806,400	Public	0.8	7008		—
Double Butte Disposal Site	Winchester	Planned	1,977,463	Public	0.8	7008		—
Earlimart Disposal Site	Earlimart	Potential	782,400	Private	0.8	7008		—
Eastern Regional Landfill	Truckee	Potential		Public	0.8	7008		—

(Continued)

TABLE 11-6   Potential Landfill Gas (LFG)-to-Energy Projects in California—Cont'd

Landfill Name	City	Current Status	MSW in Place (tons)	Landfill Owner Type	Power Generation Capacity		Medium-BTU Capacity	
					MW	MWh /Yr	MM scfd	MMBTU /Yr
Edom Hill Disposal Site	Cathedral City	Planned	4,316,700	Public	1.6	14,016		—
El Sobrante Landfill	Corona	Potential	4,900,000	Private	3.0	26,280		—
Fink Road Landfill	Crows Landing	Potential	1,142,398	Public	0.8	7008		—
Foothills Landfill	Bellota	Planned	3,766,595	Public	1.6	14,016		—
Frank R. Bowerman Landfill	Irvine	Planned	12,900,000	Public	4.8	42,048		—
Geer Road Landfill	Modesto	Potential	4,300,000	Public	2.4	21,024		—
Glenn County Landfill Site	Artois	Potential	540,000	Public	0.8	7008		—
Hanford Landfill	Hanford	Potential	1,668,000	Public	1.2	10,512		—
Highgrove Landfill	Highgrove	Planned	4,000,000	Public	1.6	14,016		—
Highway 59	Merced	Potential	4,200,000	Public	2.4	21,024		—
Imperial Landfill	Imperial	Potential	802,060	Private	0.8	7008		—
Keller Canyon Landfill	Pittsburg	Planned		Private	1.6	14,016		—
Lamb Canyon Disposal Site	Beaumont	Planned	2,279,000	Public	0.8	7008		—
McCourtney Landfill	Grass Valley	Potential	2,182,268	Public	0.8	7008		—
McFarland-Delano Landfill	Delano	Potential	1,071,000	Public	0.8	7008		—
Mead Valley Disposal Site	Perris	Planned	2,312,000	Public	0.8	7008		—
Mid-Valley Landfill	Rialto	Operational	6,200,000	Public	6.0	52,560		—
Milliken Landfill	Ontario	Operational	13,800,000	Public	8.0	70,080		—

(Continued)

TABLE 11-6   Potential Landfill Gas (LFG)-to-Energy Projects in California—Cont'd

Landfill Name	City	Current Status	MSW in Place (tons)	Landfill Owner Type	Power Generation Capacity		Medium-BTU Capacity	
					MW	MWh /Yr	MM scfd	MMBTU /Yr
Mountain View	Mountain View	Potential	11,800,000	Public	3.8	33,288		—
Newby Island Landfill– Phase II	Milpitas	Planned	19,900,000	Private	9.0	78,840		—
North County Landfill	Stockton	Potential	694,530	Public	0.8	7008		—
Olinda Landfill– Phase II	Brea	Potential	32,981,156	Public	4.0	35,040		—
Ox Mountain Landfill	Half Moon Bay	Planned	10,000,000	Private	10.0	87,600		—
Potrero Hills Landfill	Fairfield	Planned	2,538,000	Private	3.0	26,280		—
Redwood Landfill	Novato	Potential		Private	3.0	26,280		—
Ridgecrest Landfill	Ridgecrest	Potential	1,300,000	Public	0.8	7008		—
San Timoteo	Redlands	Operational	2,300,000	Public	2.0	17,520		—
Santiago Canyon	Irvine	Potential	14,275,570	Public	5.0	43,800		—
Savage Canyon Landfill	Whittier	Potential		Public	0.8	7008		—
Shafter- Wasco Landfill	Shafter	Potential	1,400,000	Public	0.8	7008		—
South Chollas Landfill	San Diego	Planned	4,750,000	Public	0.8	7008		—
Southeast Regional	Selma	Potential			0.8	7008		—
Sunshine Canyon	Sylmar	Planned	27,000,000	Private	9.0	78,840		—
Taft Landfill	Taft	Potential	1,261,106	Public	0.8	7008		—
Teapot Dome Disposal Site	Porterville	Planned	1,614,000	Public	0.8	7008		—
Tri-Cities Landfill	Fremont	Potential	7,915,544	Private	6.0	52,560		—

(Continued)

TABLE 11-6 Potential Landfill Gas (LFG)-to-Energy Projects in California—Cont'd

Landfill Name	City	Current Status	MSW in Place (tons)	Landfill Owner Type	Power Generation Capacity		Medium-BTU Capacity	
					MW	MWh /Yr	MM scfd	MMBTU /Yr
Tuolumne County Central Landfill	Jamestown	Potential	986,400	Public	0.8	7008		—
Union Mine Disposal Site	El Dorado	Potential	1,900,000	Public	0.8	7008		—
Upper Valley Disposal	Napa	Potential	570,000	Private	0.8	7008		—
Vasco Road Landfill	Livermore	Planned	15,000,000	Private	4.8	42,048		—
Victorville Landfill	Victorville	Potential	600,000	Public	0.8	7008		—
West Central Landfill	Redding	Potential	632,098	Public	1.6	14,016		—
Yuba-Sutter	Marysville	Potential	2,268,000	Private	1.2	10,512		—
					177.9	1,558,404		

MSW = municipal solid waste; MMscfd = million standard cubic feet per day.
(Source: SCS Engineers. 2002. *Economic and financial aspects of landfill gas to energy project development in California*. The California Energy Commission. Sacramento, CA.)

enclosed areas, where methane concentrations in excess of methane's lower explosive limit (5%) can accumulate. The risk of explosion is not limited to structures on the landfill, since LFG can migrate hundreds of feet under ground beyond a landfill boundary. Methane migration represents a risk to existing structures and an obstacle to new commercial and residential property development. The risk is a long-term risk since most landfills will continue to generate significant quantities of LFG for more than 30 years after closure. A properly designed LFG collection and control system can mitigate this risk; in fact, dozens of golf courses, office parks, convention centers, and residential developments have been built on or immediately adjacent to closed landfills after implementation of proper LFG control. In some cases, these facilities have become users of LFG.

Environmental Issues.   The principal component in LFG is methane. Methane is believed to have more than 20 times the impact as a GHG than CO_2 on a weight basis. Collection and destruction of this methane, typically in a flare, converts the methane into CO_2, which greatly reduces the greenhouse impact of LFG. If the LFG is used as a fuel, it displaces combustion of another fuel and generates an additional offset.

A landfill with 5.5 million tons of typical refuse in place will emit in the range of 300,000 tons per year of CO_2 into the atmosphere. Collection and utilization of LFG can make significant national and worldwide contributions to GHG-control efforts.

LFG is odorous and can be a nuisance to neighbors. LFG contains low levels of volatile organic compounds and toxic organic compounds (including toluene, xylene, and benzene). While present in low levels, significant quantities of volatile organic compounds and toxic compounds can be emitted to the atmosphere, due to the large volumes of LFG which landfills produce.

A 5.5 million ton landfill can emit in the range of 83 tons per year of volatile organic compounds and 55 tons per year of toxic organic compounds. Volatile organic compounds are a precursor to ground-level ozone formation. LFG emissions are a threat to local air quality, and hence to public health. The volatile organic, toxic, and odorous compounds present in LFG are, however, readily destructible (typically 98%+) through combustion in flares or in conventional power-generation and fuel-burning equipment.

Energy Value.   The principal benefit of LFG collection, from the perspective of the LFGTE industry, is that its collection produces a low-cost, fairly clean fuel. If an LFG-collection system is installed for the purpose of achieving the above-outlined environmental and property protection benefits, then the fuel is availab le at a flare station at no net cost. Even if the total cost of LFG collection is allocated against its energy value, the cost of production (capital and operations cost) is usually much lower than the cost of natural gas. Often, the landfill owner and the fuel user strike a cost-sharing agreement covering the cost of well field installation and operation.

Pathways to Project Implementation

Self-Development Versus Second-Party Developers.   The most significant contractual/financial decision to be made by a landfill owner in the development of an LFGTE project is the decision whether to self-develop the project or to turn the project over to an independent developer. An independent developer is sometimes called a "second" party developer. The landfill owner is the "first" party. The landfill owner must weigh two considerations when making this decision:

- The landfill owner's willingness to accept project risk
- The landfill owner's willingness to commit time and attention away from his or her principal business

The landfill owner's willingness to accept project risk is the most important consideration in the decision-making process. If the landfill owner elects to self-develop a project, then he must provide the capital. The capital is 100% at risk. He receives 100% of the project benefits, but is also exposed to 100% of the project losses. A second party will provide the capital for the project and insulate the landfill owner from project risk. The second party will compensate the landfill owner for the right to develop a project at his landfill; however, this compensation will be limited to only a fraction of the economic benefit produced by the project.

If the landfill owner decides to self-develop a project, the landfill owner can minimize the demand on his and his staff's time by hiring consultants to handle almost all of the details of project development and implementation; however, heavy reliance on consultants will result in additional funds being placed at risk in the earlier stages of project development. The heavy use of consultants reduces the concern with respect to the issue of time/attention diversion, but trades this benefit off against an increase in financial risk.

Project Risks.   The two sections that follow discuss project risk and how project risks can be shared or minimized. A full understanding of project risk is necessary to make an informed decision on the preferred method of project development. Methods for mitigating and distributing project risk are also discussed. The risks will be discussed in the context of an electric-power project since almost all of the LFGTE projects in California are or will be electric-power projects.

An LFGTE electric-power project, like any independent power project, must face and successfully overcome several uncertainties. The principal uncertainties faced by an independent power project include the following:

- Total project construction cost
- Security of fuel supply and price
- Non-fuel annual operating cost
- Environmental and other permitting
- Plant performance
- Ability to secure financing
- Ability to complete on schedule

Total project construction cost includes the following:

- The initial costs incurred in project development (i.e., feasibility studies, permitting, legal/administrative costs of securing power sales and other agreements)
- Financing costs (i.e., loan initiation costs and interest during construction)
- Construction costs through final change orders
- Initial funding of operating costs (during shakedown and start-up until electric power and its revenues are reliably generated)

The project risk associated with the uncertainty of project construction cost is, of course, that project construction cost may be underestimated, resulting in a decrease in the net income generated by the project. Project construction costs can increase due to a poor initial cost estimate or due to unforeseeable factors. Unforeseeable factors can include the following:

- Permitting difficulties resulting in schedule delays and their associated cost escalation
- Cost increases due to hardware additions to cover more stringent environmental controls, including advanced air emission controls and noise controls
- Increased consultant costs for permitting

As the project develops beyond its initial stages and as more money and effort are expended, total project cost becomes better confirmed. A project can be terminated at any point; however, monies spent to the point of termination are lost. The major milestones in confirmation of total project construction cost are receipt of relevant permits, with identification of their attendant construction cost impacts, and execution of a firm construction cost contract.

Security of fuel supply and price is an important consideration in an independent power project since the projects are often financed over a 10- to 20-year period. While a commitment to supply coal or natural gas over a long term can fairly easily be secured, the unit price for these commodities is often difficult to fix or cap in future years. LFG has an advantage over fossil fuels in that (1) if a project is self-developed, the developer already owns the fuel; and (2) if there is a second-party developer, then the LFG price can be agreed upon with certainty for the entire term of its anticipated use. The price need not be established on a strict dollar-per-MMBTU basis, but could be set as a percentage of the gross power revenue generated by the project or on another mutually agreeable basis. While the certainty of price is better for LFG than for fossil fuel, its position with respect to certainty of supply is not as favorable. LFG availability is based on projections which incorporate assumptions on the waste's gas generation potential and on the stability of long-term landfill operations. Questions and concerns include the following:

- Will the landfill stay open for its permitted life and continue to receive wastes at the projected quantities? Regulatory changes, public pressures, and changing market conditions can affect landfill life and waste receipt. Re-permitting or permitting of site extensions might be particularly troublesome.
- Will the landfill owner vigilantly operate the gas-collection system to maintain a high quality and quantity of gas flow? In most cases, gas-collection systems were installed to reduce air emissions, control odors, or mitigate safety problems, and not to produce fuel gas. Further, day-to-day landfill operations can result in intermittent disruptions in gas supply.
- How accurate is the projection of gas production? While LFG-recovery models have improved in recent years, they must often rely on imperfect input information. If a complete LFG-collection system has been installed on at least a large segment of the existing landfill, it is possible to site calibrate the model, reducing much of the uncertainty in the LFG-recovery projection.

One drawback to LFG is that it is available only where it is produced. The power plant must be sited at or fairly close to the landfill. Landfills are often located in regions where air emissions permits are difficult to obtain. Finally, the project must be sized based on what LFG is available, not on what size makes the most economic/technical sense.

The uncertainty of non-fuel annual operating on independent power projects is not great. This is also true for LFGTE projects. Initial-year labor, routine maintenance, and commodity consumption can be projected with a fair amount of certainty and the escalation of these costs can be established using reasonable judgment. The largest uncertainty is probably the extent of "non-routine" repair/replacement costs. Major equipment failures as a result of operator error or equipment defects beyond the warranty period are examples of non-routine repair/replacement costs. The hundreds of operating LFGTE power projects in the United States and the dozens of projects in California provide a database for operating cost information.

Environmental and other permitting is an area of uncertainty which is faced by virtually every commercial or public venture undertaken which involves construction. The principal concerns with permitting are cost impact and schedule impact. The schedule question, simply stated, is: How long will it take to secure the necessary permits and, in fact, can the more sensitive permits be secured with any reasonable expenditure of time and money? The conditions contained in the ultimately issued permits which govern air emissions, noise control, condensate disposal, and ongoing environmental monitoring can greatly affect project construction and operating cost. As compared to other independent power projects, LFGTE projects are generally at an advantage with respect to permitting since they often result in a net reduction in air emissions and are sited in an area somewhat buffered from the public. In some instances, an LFGTE power project can be permitted as a routine matter.

The principal measures of plant performance are as follows:

- Heat rate (fuel consumption/kWh produced)
- Plant output (kW)
- Air emissions
- Availability (ratio of the time the plant is able to operate to the total hours in a year)

Uncertainties associated with heat rate and plant output are not great when using proven technologies and experienced designers/equipment manufacturers. The same statement can generally be made with respect to air emissions; however, as technology is pushed to achieve lower levels of emissions, then the ability to comfortably achieve mandated air emissions becomes a concern. Most of the risks associated with heat rate, plant output, and air emissions can be passed on to the equipment manufacturer and/or construction contractor; however, performance guarantees from equipment manufacturers and contractors normally involve commitments to simply rectify the deficiency and not to reimburse the owner for periods of reduced power output while the remedy is being implemented. Equipment manufacturers and contractors cannot bear the cost of consequential damages within their normal profit margins.

The issue of availability is one measure of performance that cannot be guaranteed over the long run by an equipment manufacturer or contractor—unless he operates and/or maintains the plant at some premium cost. If the project owner desires comfort in this area from the equipment manufacturer and/or contractor, he must normally rely on short-term availability tests. Such tests are typically in the form of 7- to 30-day reliability tests at contract closeout. In the case of LFGTE power projects, the equipment which is used has a long and successful track record in LFG service. Developers normally assume that their project will operate at availability levels equal to similar LFGTE projects.

The risk of financing is whether funds can be raised and whether they can be secured at reasonable terms. In a privately financed project, this uncertainty is overcome by demonstrating that the project will, with little doubt, produce a reasonable rate of return on investment and will, at a minimum, generate sufficient revenues to cover operating costs and debt service given the development of reasonably likely adverse conditions. Adverse conditions might be lower than forecast availability (a slightly greater prospect on an LFG project than a natural gas–fired project) or fuel costs escalating more rapidly than assumed in the life cycle economic model (much less likely than for a natural gas–fired project). The life cycle cost model is normally called a *pro forma* balance sheet in the independent power industry. In a publicly financed project, the test is sometimes less stringent and the criterion is sometimes that the project simply does better than "break even." In a private non-recourse financing project, the risk of financing is not eliminated until all other risks have been minimized, and this usually occurs only after a construction contract is in place. The availability of financing and its terms on a public project can be confirmed much earlier if so desired. A private firm with a strong balance sheet and several successful, operating projects is generally in the position that it knows that it can secure financing. Project financing is discussed in greater detail in later sections of this report.

The final uncertainty is the project completion schedule—more specifically, how long will it take before the project is on the grid producing power? The above discussion identifies two major schedule concerns—how long will it take to secure permits, and will there be problems in start-up? Delays in initiation of construction and in commercial operation result in cost escalation and postponement of anticipated revenue streams which will not only reduce the profitability of a project, but could also subject the project owner to cash flow problems. Some power sale agreements or incentive packages have "sunset" dates at which point the agreement is subject to cancellation or renegotiation and/or call for meeting of certain "milestones." Failure to meet the milestones can result in cancellation of the agreement or incentive.

There are two general rules with respect to risk on independent power projects which are even more applicable to LFGTE power projects:

- Rule 1: Risk varies in indirect proportion to money and time expended.
- Rule 2: When multiple parties are involved in the execution of a project, the net profit which it generates must be shared in general portion to the risk allocated to and contribution made by each party.

As shown in Figure 11-18, uncertainty is reduced in a stepwise manner as specific milestones are met. The cumulative amount of money committed to the project also increases with each milestone. If an individual milestone causes the project to be considered infeasible, then the monies spent to that point are lost. If this occurs early, say at the completion of the feasibility study, then only the cost of the study is lost. If this occurs at the point of financing, then the cost of all work to that point is lost.

In general, the risks associated with an LFGTE power project are no greater than, and in some areas are less than, those associated with a conventional power plant.

Risk Allocation.   An LFGTE power project can have a number of participants, including the following:

- The landfill owner
- The owner of the gas-collection facilities (if other than the landfill owner)
- The operator of the gas-collection facilities (if other than the landfill owner)
- The owner of the power plant
- The operator of the power plant
- The construction contractor
- The manufacturers of the major equipment
- The architect/engineer
- The equity holder(s)
- The holder(s) of long-term debt

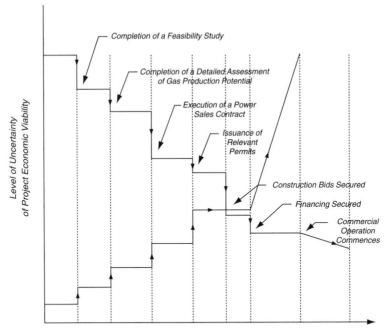

FIGURE 11-18 *Development risk as a function of time and money. (Source: SCS Engineers. 2002. Economic and financial aspects of landfill gas to energy project development in California. The California Energy Commission. Sacramento, CA.)*

In many project structures, one firm or entity fulfills several of these roles. Risk can be spread among these parties, depending on their interest in the project. The holder(s) of the long-term debt are perhaps the most risk averse. They will seek to mitigate their risk by looking to the project developer to guarantee loan repayment through revenue sources other than the project itself and through the commitment of collateral other than the project itself. If the borrowing is on a "non-recourse" basis, the lenders will expect all other project participants to absorb virtually all risk. The holder(s) of long-term debt look at the project as an opportunity to lend funds at an interest premium, but want virtually 100% assurance that their investment will be repaid. The equity holder(s) are more entrepreneurial and bear the bulk of the financing risk. For accepting this risk, they will expect to be rewarded fairly handsomely—typically an internal rate of return in the range of 20% to 25%. The bulk of the equity is normally held by the developer and his usual investment partners or, when the landfill owner acts as the owner-developer, the landfill owner himself. The equity holders, like the long-term lender(s), will attempt to pass risk along to other project participants. Project financing is discussed in greater detail in later sections of this chapter.

Typical strategies for risk transfer to various parties include transfer of risk to the following:

- Equipment manufacturers: By seeking assurances on equipment performance and delivery time in the form of monetary penalties and other guarantees.
- Construction contractor: By seeking an overall wrap-around guarantee on a total project basis in the areas of performance and schedule. A wrap-around is more easily obtained from a turnkey contractor rather than a general contractor.
- Architect/engineer: By seeking guarantees on those aspects of schedule and performance under his or her direct control. Architect/engineers have traditionally limited their warranty to the redesign of deficient work; however, they have been increasingly accepting some risk, particularly when subcontracting to turnkey contractors.
- Landfill operator (or gas producer): By seeking discounts on fuel payments based on problems with gas delivery and gas quality.
- Power plant operator: By seeking bonus/penalty arrangements from the operations contractor keyed to plant availability (power output) and cost control.

Alternative Overall Implementation Structures.   Figure 11-19 identifies several alternative structures for implementing LFG power projects. The alternatives vary to some degree depending on whether the landfill owner is a public or private entity or whether a decision is made to self-develop or to employ an independent developer in implementing the power-generation project. The implementation options vary in their structure primarily in who installs/owns/operates the gas-collection facilities and who installs/owns/operates the power plant. The project would normally involve only one or two parties—two parties when a developer is employed and one party when the landfill owner self-develops. The qualification for tax credits usually results in a second or third party being brought into the project structure. The existing federal alternative energy production tax credit is only available if the gas which is produced is sold to an "unrelated party" and is only of benefit if the producer has an "appetite" for tax credits. The second or third party must not necessarily be totally unrelated to the other parties, and structures have been developed which involve the principals in ways acceptable to the Internal Revenue Service. Figure 11-19 does not show every possible implementation structure, but it does show the most likely arrangements. In all options involving a second party, the landfill owner (either public or private) receives compensation in some form. The compensation can take the form of any of the following:

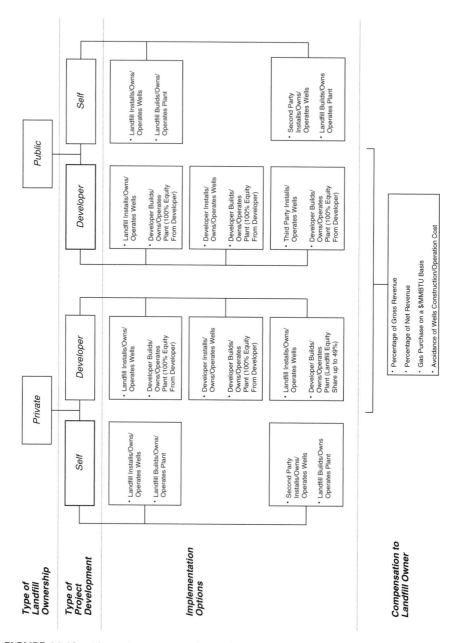

FIGURE 11-19 *Alternative structures for implementing landfill gas power projects. (Source: SCS Engineers. 2002.* Economic and financial aspects of landfill gas to energy project development in California. *The California Energy Commission. Sacramento, CA.)*

- A percentage of gross power revenue
- A percentage of net revenue
- Gas purchase on a $/MMBTU basis
- The avoidance (or reduction) in the cost of well field construction and operation/maintenance cost (if the landfill owner needed to install the well field for environmental or regulatory purposes)

Pertinent Regulations

California Code of Regulations (CCR) Title 27 requires that methane concentrations in the soil at the property boundaries of landfills not exceed 5%. The purpose of this regulation is to prevent the off-site migration of LFG in explosive concentrations. LFG collection and control systems in California are often installed to satisfy CCR Title 27. Responsibility for administering CCR Title 27 is delegated by state government to county or city government. The agencies responsible for enforcing CCR Title 27 are commonly known as local enforcement agencies.

The U.S. EPA's New Source Performance Standards (NSPS) for municipal solid waste landfills require that LFG collection and control systems be installed at all landfills which have a design capacity over 2.5 million Mg (2.75 million tons) and are projected to emit more than 50 Mg (55 tons) of NMOCs to the atmosphere per year. In California, a landfill with as little as 1 million tons of waste in place could be required to install an LFG collection and control system; however, a landfill with as much as 5 million tons of waste in place may not be required to install an LFG collection and control system. The authority for administering NSPS has been delegated to the individual air-quality management districts (AQMDs) and air-pollution control districts (APCDs) throughout the state. NSPS was promulgated in 1996. Many of California's AQMDs and APCDs already had preexisting LFG collection and control rules. Some of these rules were and continue to be more stringent than NSPS. In some cases, these rules require that landfills having less than 1 million tons of waste in place install LFG collection and control systems.

The benefit of CCR Title 27, NSPS, and related rules, from the perspective of the LFGTE industry, is that these rules require that LFG collection and control systems be installed solely to satisfy a regulatory purpose. The LFG is available at no cost. There are very few landfills in California which could support an LFGTE project greater than 0.5 MW in size which are not required by environmental regulations to install an LFG collection and control system.

NSPS requires that once LFG is collected, it must be processed through a device capable of achieving a 98% destruction of NMOCs (or an exhaust concentration of 20 ppmv). An enclosed flare is normally employed for LFG destruction. When LFG is used beneficially, the LFGTE facility replaces the flare for the portion of LFG which is beneficially used. LFGTE facilities, at NSPS sites, must also achieve a 98% destruction of NMOCs. If, for a particular technology or landfill, 98% destruction exceeds what would be Best Available Control Technology (BACT) from the perspective of the power plant alone, then the 98% destruction would still govern. If LFG is "treated," as it is in a pipeline-quality project, then the LFGTE project is exempt from the NSPS requirement. It is unclear at this point whether a medium-BTU project would need to satisfy this requirement. A medium-BTU project only subjects the LFG to light cleanup, and the U.S. EPA has not issued clear guidance as to whether or not a medium-BTU project would satisfy their definition of treatment. At the present time, this issue needs to be addressed on a case-by-case basis during permitting.

An LFGTE project, like any project which generates air emissions, must obtain an air permit from an AQMD or APCD having jurisdiction over the landfill. The permitting processes typically involve determination of BACT, addressing offset issues, and health risk assessments. The air permitting process for an LFGTE project is generally no more difficult than for a project employing conventional fuel. Most AQMDs and APCDs are very supportive of LFGTE projects.

Impediments to LFGTE Development

Environmental/Permitting Issues. Environmental and permitting issues seldom represent impediments to LFGTE development. LFGTE projects are normally located at a landfill, and the installation of a relatively small energy-recovery facility at a site already committed to waste disposal is not viewed as a contradictory landfill use. The environmental impacts are minor. Air emissions are partially offset by reduced flaring of LFG. In a few instances, establishment of BACT and securing arrangements for offsets have delayed LFGTE project implementation; however, this is a very AQMD/APCD-specific problem. For the most part, the determination of BACT has been straightforward and offsets have not been an issue.

Interconnection.   LFGTE power projects are by definition qualifying facilities because they are fired on waste fuel. As a qualifying facility, a power project is guaranteed the right to interconnect to the utility's grid at reasonable costs. Most LFGTE power plants are relatively small and the impact on the utility's grid is not significant. In some cases, however, the utility must increase the capacity of its transmission lines between the location of the LFGTE project and the location on the grid that can accommodate the amount of power produced by an LFGTE power plant. Under such circumstances, the interconnection can become a significant part of the plant's total construction cost. Impact on project schedule can also be a concern. The larger the impact on the utility, the longer the lead time needed by the utility to schedule and complete the interconnection work.

Interconnection problems can be avoided by contacting the local utility in the early stages of project development. Early identification of interconnection requirements will mitigate schedule delays and will allow the costs of interconnection to be fully reflected in the financing of the project.

California investor-owned utilities (IOUs) have their interconnection requirements specified in their Rule 21, which is filed with the California Public Utility Commission (CPUC). Implementation of Rule 21 requirements is fairly uniform for the California IOUs. An interconnection application, satisfying the requirements of Rule 21, must be filed with the local utility to begin the interconnection evaluation process. Municipal utilities have their own requirements governing interconnection, and the requirements of each utility must be individually addressed.

Economics and Financing.   Individual LFGTE projects require only small to moderately large capital investments. Due diligence, legal, and administrative activities and costs, which are associated with securing private debt and equity, are largely unrelated to the amount of money involved. It is sometimes difficult to attract private investment when a capital investment is relatively small. In addition, there are a very limited number of lenders and investors who have LFGTE experience.

The issue of investment size can often be overcome by aggregating individual projects into a portfolio containing several projects.

LFGTE Economics

Capital and Operating Costs of LFG Collection and Control in California

Conventional Landfills.   The principal components of an LFG-collection system are its extraction wells, the LFG-collection piping (which allows the LFG to be drawn to a central location), and a blower/flare station (which contains vacuum blowers to pull the LFG to the blower/flare station and a flare to burn the LFG).

The extraction wells can consist of horizontal wells (known as *trench collectors* or *horizontal collectors*) or vertical wells. Horizontal wells can only be installed as waste is being placed. Vertical wells can be installed as waste is being placed or after a section of a landfill (or an entire landfill) is closed. The LFG-collection piping can be buried, laid on the surface, or placed on pipe supports. The installed cost of an LFG-collection system varies based on the type of system installed. The cost is also affected by the depth of the waste in the landfill. The above variables make it difficult to quote general construction costs. A typical well field construction cost is $15,000 per acre (with a range of $10,000 to $20,000 per acre). A typical well field operation/maintenance cost is $650 per acre (with a range of $400 to $900 per acre) per year.

The construction cost of a blower/flare station is a function of LFG flow rate. LFG flow rate is primarily a function of the amount of waste in place. The amount of precipitation at the landfill and the average age of the waste are important factors affecting LFG flow rate. As a result, costs are best estimated on a per-scfm basis. Blower/flare station construction costs range from $350 to $450 per scfm. Operation/maintenance costs range from $20 to $30 per scfm per year.

The total cost of LFG recovery (on an energy basis), assuming retirement of capital cost over 10 years at in interest rate of 10%, is about $0.90/MMBTU (with a range of $0.60 to $1.20/MMBTU).

As discussed in the Landfill Gas–to–Energy Overview section, most landfills in California having significant LFG production potential are required by regulation to install LFG-collection systems. As a result, there is no cost assignable to the energy recovered. LFG is sold to second parties or is used in self-developed projects to defray part of the cost of construction and operation/maintenance of the LFG-collection facilities. When sold to second parties, the cost of LFG is a value set by market conditions, principally based on the value of the end product sold (e.g., electric power or medium-BTU gas), rather than on the actual cost of LFG production.

Bioreactor Landfills.   At the present time, bioreactor landfills are in the early stages of full-scale field testing. The construction and operation/maintenance costs associated with bioreactor landfills are not known. Construction and operation/maintenance costs will be much higher for bioreactors (versus conventional landfills) on a per-acre basis; however, the waste management industry expects lower long-term life cycle costs on a dollar-per-ton basis. Even if costs were well defined, there is no clear method to allocate costs between LFG recovery and other bioreactor components. It is likely that LFG-recovery costs on a dollar-per-MMBTU basis will be lower for a bioreactor landfill than for a conventional landfill. The cost reduction could be in the 25%-to-50% range, depending on how bioreactor costs are allocated.

Costs of Medium-BTU Projects.   Direct-use (medium-BTU) projects have three components: the compressor plant; the pipeline to deliver the LFG to the end user; and end-user modifications to support LFG firing or LFG co-firing. The cost of the compressor plant is a function of its design flow rate and design pressure. The design pressure required from one project to another does not greatly vary and is typically in the range of 50 to 100 psig. Construction costs and operation/maintenance costs are primarily a function of flow rate. Flow rate is typically quoted in million standard cubic feet per day (mmscfd). The total installed cost of a medium-BTU compressor plant is in the range of $600,000 to $700,000/mmscfd. Operation/maintenance costs are about $400 to $500/mmscf (assuming electric motor drives at current California power costs). LFG can be processed at a cost of about $600 to $800/mmscf (or $1.20 to $1.60/MMBTU), assuming retirement of capital cost at an interest rate of 10% over 15 years.

The cost of the pipeline to an end user is a function of its length and its diameter. The diameter is a function of the design flow rate. Pipeline length (versus diameter) accounts for more than 80% of the pipeline cost. Pipeline cost is in the range of $30 to $40 per foot. Pipeline operation/maintenance costs are insignificant. Each 5000 feet of pipeline adds about $0.15/MMBTU to the LFG delivery cost.

The cost of end-user modifications is highly variable, but these costs are usually borne by the end user and paid for through the cost savings he experiences from the use of LFG. The cost of LFG to the end user is generally indexed to his avoided conventional fuel cost. The discount which must be offered against the end user's avoided cost is partially a function of his expected conversion cost.

Capital and Operating Costs of Power-Generation Facilities in California

Reciprocating Engines.   The total installed cost of an LFG-fired reciprocating engine power plant is in the range of $1100 to $1300/kW. The scope of the installation would begin with an LFG booster blower and end with a step-up transformer. The variability in price relates to differences in plant size, site conditions, and whether or not the equipment is installed in a building or is supplied in its factory-shipped containers. Larger plants and containerized installations would fall in the lower end of the price range. The price range quoted assumes a minimum plant size of 800 kW. The cost per kW does not significantly decrease when the plant size increases above 6 MW.

The operation/maintenance cost of an LFG-fired reciprocating engine power plant, exclusive of LFG-recovery cost, is in the range of $0.016 to $0.020 per kWh. The lower cost is associated with larger plants (3 to 6 MW) and the higher cost is associated with smaller plants.

The total cost of power production, based on net power output and assuming retirement of the capital cost over 15 years at an interest rate of 10%, would be in the range of $0.035 to $0.053 per kWh.

Combustion Turbines.   The total installed cost of an LFG-fired combustion turbine power plant is in the range of $1000 to $1200/kW. The scope of the installation would begin with an LFG compressor and end with a step-up transformer. The price range quoted above assumes a minimum plant size of 3.5 MW. Smaller LFG-fired combustion turbine plants are generally no longer being built and, if one were built, the cost would exceed $1200/kW.

The operation/maintenance cost of an LFG-fired combustion turbine power plant, exclusive of LFG-recovery cost, is in the range of $0.014 to $0.018 per kWh. The highest cost is associated with the smallest plants (3.5 MW), and the lowest cost is associated with larger plants (10 MW+).

The total cost of power production, based on net power output and assuming retirement of the capital cost over 15 years at an interest rate of 10%, would be in the range of $0.034 to $0.042 per kWh.

The cost of power production with a combustion turbine plant is comparable to the cost of a reciprocating engine plant, when similar in size, despite the combustion turbine's lower capital and operation/maintenance costs, because the ratio of net power output to gross power output is much lower for a combustion turbine power plant than for a reciprocating engine power plant. This causes net cost on a cent–per–kilowatt-hour basis to increase.

Steam-Cycle Power Plants.   The total installed cost of a steam-cycle power plant (on a dollar-per-kilowatt basis) varies greatly with plant size. It is generally acknowledged that steam-cycle power plants are not cost competitive with other LFG power-generation technologies at capacities under 20 MW. Plants smaller than 20 MW in size have been built; however, selection of a steam cycle in such instances has generally been due to air emissions issues, the availability of used equipment, or an owner preference for lower long-term operation/maintenance costs at the expense of higher initial capital costs.

The total installed cost of an LFG-fired steam-cycle power plant is in the range of $2500/kW (at 10 MW) to $1500/kW (at 30 MW+). The scope of the installation would begin with an LFG booster blower and end with a step-up transformer.

The operation/maintenance cost of an LFG-fired steam-cycle power plant, exclusive of LFG-recovery cost, is in the range of $0.010 to $0.014 per kWh.

The total cost of power production, based on net power output and assuming retirement of the capital cost over 15 years at an interest rate of 10%, would be in the range of $0.036 to $0.055 per kWh.

Microturbines. The total installed cost of an LFG-fired microturbine power plant can range from $1800 to $3000/kW. For a 30-kW installation, the cost is in the range of $2500 to $3000/kW. The cost of a 300-kW installation is in the range of $1800 to $2100/kW. Above 300 kW, there is little reduction in cost on a dollar-per-kilowatt basis. Further cost reductions are not possible since the maximum currently available microturbine size is 75 kW. When and if larger microturbines become available, then the cost of a microturbine installation over 300 kW will decrease on a dollar-per-kilowatt basis.

The operation/maintenance cost of an LFG-fired microturbine power plant, exclusive of LFG-recovery cost, is in the range of $0.018 to $0.022 per kWh.

The total cost of power production, based on net power output and assuming retirement of the capital cost over 10 years at an interest rate of 10%, would be in the range of $0.065 to $0.093 per kWh.

Fuel Cells. The total installed cost of a fuel cell using LFG as feedstock would be in the range of $3800 to $4000/kW. The fuel cell which is currently commercially available for LFG service costs $600,000 or $3000/kW. The operation/maintenance cost for an LFG fuel cell facility would be in the range of $0.020 to $0.025 per kWh.

The total cost of power production, based on net power output and assuming retirement of capital cost over 15 years at an interest rate of 10%, is in the range of $0.09 to $0.10 per kWh. Fuel cells employing waste material as feedstock often qualify for significant capital cost subsidies; however, even with a capital cost subsidy of 50%, fuel cells are currently not cost competitive with other LFG power-generation technologies.

Costs of Pipeline-Quality Gas Production Projects

Pipeline-quality gas projects are generally in the 5 to 10 mmscfd (inlet flow) size range. A typical installed cost is $1.2 million to $1.5 million per mmscfd (inlet). Typical operation/maintenance costs are $0.50 to $1.00/mcf (outlet). Pipeline-quality gas can be produced at a cost of $1.70 to $2.20/mcf (of natural gas equivalent), assuming retirement of capital cost over 15 years at a 10% rate of interest.

Total Cost of LFGTE Systems

The total construction costs, operation/maintenance costs, and net production costs of an LFGTE system can be roughly estimated by use of the above outlined cost-estimating guidelines. The costs of LFG collection can be included or excluded in these estimates, depending on the view taken toward these costs. The net production cost of electric power production using LFG varies greatly depending on project size and technology. The cost can range from as low as $0.034 per kWh to as high as $0.10 per kWh.

Medium-BTU gas can be delivered to an end user for a cost in the range of $1.70 to $2.10/MMBTU, assuming that a 3-mile pipeline is required.

Pipeline-quality gas can be produced for $1.70 to $2.20/mcf.

The above costs do not consider the beneficial impact of tax credits or other incentives. What follows is a discussion of tax credits and incentives which are available to aid LFG-recovery projects.

Financing of LFGTE Projects in California

Incentives and Credits Available to Support LFGTE Development

Federal

Section 29 Tax Credit.   Section 29 of the Internal Revenue Code provides a tax credit to support the use of fuel from specific non-conventional sources. LFG, considered to be a biomass fuel, qualifies for this tax credit. The Section 29 tax credit is, and continues to be, the major incentive assisting the LFG utilization industry. The Section 29 tax credit is currently valued at about $1.10 per MMBTU.

Under current Internal Revenue Service regulations, Section 29 tax credits can be claimed through 2007 for LFG-collection facilities placed in service after December 31, 1992, and prior to June 30, 1998. Facilities installed after December 31, 1996, and prior to June 30, 1998, had to be constructed under a binding contract executed prior to December 31, 1996. LFG-collection facilities placed in service prior to December 31, 1992, could claim Section 29 tax credits only through 2002.

It is important to note that it is not necessary for the LFG beneficial use to have commenced operation before June 30, 1998. It is only necessary that the well field be placed in service by that date. The beneficial use can come online at any time through the diversion of LFG from a flare to a beneficial use. From the point of diversion forward, tax credits can be taken. A "grandfathered" well field is all that is necessary to claim tax credits.

The Section 29 tax credit is taken by the fuel producer. The fuel producer is the owner/operator of the well field. The LFG must be sold to an "unrelated party" for beneficial use. These conditions, coupled with the fact that all qualifying well fields are pre-existing, requires that innovative contractual arrangements be employed to implement tax credit transactions. It is very common for well field owners, typically landfill owners, to sell or lease their LFG-collection system to parties better situated to make use of the tax credits. After well field ownership has transferred, the "unrelated party" test can still be met by the project developer by implementing the project through two independent companies—a gas-producing company (Gasco) and a power-generating company (Genco).

The Internal Revenue Service periodically issues Private Letter Rulings (PLRs) which change and clarify the definition of *facilities* and the definition of *replacement* wells. LFGTE developers and their tax consultants must carefully review these PLRs as they often change the qualifying criteria for well fields upgraded or modified after 1992 and 1998.

Maximum value can be obtained from what remains of the Section 29 tax credit by:

- Focusing on landfills which have 2007 well fields
- Focusing on landfills which had comprehensive pre-1998 horizontal LFG-collection system coverage (with expected refuse overfill)

The latter consideration is of value since the grandfathered LFG-extraction facilities will in most instances be capable of collecting part or all of the LFG being provided by the overfill.

Renewable Energy Production Incentive.   The Renewable Energy Production Incentive (REPI) was authorized under Section 1212 of the Energy Policy Act of 1992. The REPI is administered by the Department of Energy (DOE). The REPI provides an incentive payment of $0.015/kWh (1993 dollars) with annual increases for inflation for electric power produced by qualifying energy facilities. Qualifying facilities are those that meet all of the following conditions:

- Owned by state and local governmental entities (including municipal utilities)
- Have commenced operation between October 1, 1993, and no later than September 30, 2003
- Have a renewable energy source as a source of the electricity (LFG is a qualifying renewable energy source)

Payments are made for a 10-year period commencing with the first year claimed. The amount of money available for the REPI is established every year by congressional appropriation. If insufficient funds are available to fund all applications, the funds are divided amongst the applicants. Power generation not reimbursed in a given year can be rolled over into future years. The funding is prioritized by project type. Tier 1 projects employ solar, wind, geothermal, or closed-loop biomass technologies. Tier 2 projects include other renewable technologies including LFG. Tier 1 projects receive first priority for 100% funding.

The DOE has adopted a fairly broad definition for the term *state and local government*. The definition includes, for example, schools and public authorities. The University of California at Los Angeles has claimed REPI benefits since 1994. It is not necessary that the power-generation unit be 100% LFG fired. The fraction of the total kilowatt-hours produced on LFG can be calculated, and the REPI can be claimed on the LFG-fired fraction. There is no minimum LFG fraction required.

In 1994 and 1995, sufficient appropriations were made by Congress to fund all Tier 1 and Tier 2 projects at 100%. Since 1996, only Tier 1 projects were funded at 100%. Tier 2 projects received partial payments on a prorated basis. Full funding of REPI applications would currently require an annual appropriation level in the range of $8,000,000 to $9,000,000. Recent annual appropriations have been in the range of $1,500,000 to $4,000,000.

A new LFG-fired project now theoretically qualifies for well over $0.015/kWh. It is not possible to determine how much a new LFG-fired project will actually receive since: (1) the total funding level varies each year; (2) the number of new projects and their Tier 1 versus Tier 2 mix in a given year is unknown; and (3) it is difficult to calculate the impact of the moving inventory of partially funded kilowatt-hours from prior years. Based on recent funding levels, the REPI could probably currently contribute less than $0.005/kWh to new LFG-fired projects.

Pending Federal Initiatives.   Over the last 4 years, the LFGTE industry has vigorously pursued an extension to Section 29 or an equivalent tax credit.

As of the date of this report, several proposals were pending in Congress to extend LFGTE tax credits in some form. A tax credit in the range of $0.015/kWh (or its equivalent for projects with gas as their product) is common to several proposals.

State.   California has a comprehensive package of incentives and subsidies to aid renewable-energy electric power projects. The incentives and subsidies grew out of California's program to deregulate its electric-power industry and recent efforts to stimulate power production and/or reduce energy consumption.

The principal vehicle for assisting renewable energy projects is the Renewable Resource Trust Fund. The program is administered by the California Energy Commission (CEC). The fund has four accounts:

1. Existing Renewable Resources Account
2. New Renewable Resources Account

3. Emerging Renewable Resources Account
4. Customer-Side Renewable Resource Purchases Account

The first account provides a subsidy for existing projects when the power sales rate of existing projects falls below a specified benchmark price. The second account pays a fixed subsidy of up to $0.015/kWh for new projects (for a 5-year period). The actual price paid for a specific project is determined through submitting a winning bid for a subsidy through auctions periodically administered by the CEC. The third account will provide capital cost grants equal to the lesser of 50% or $4000/kW for specific emerging technologies. As of June 2001, the fuel cell is the only technology which could utilize LFG which could qualify for the emerging renewable resources account. Complete information on the assistance available through the Renewable Resource Trust Fund can be found on the CEC Web site (http://www.energy.ca.gov).

Other state programs which could and can provide assistance to LFGTE projects include the following:

- Self-Generation Program: The program is administered by Southern California Edison (SCE), Pacific Gas and Electric Company (PG&E), the San Diego Energy Office, and the Southern California Gas Company for cogeneration projects under 1 MW. The program provides 30% to 50% grants for projects satisfying specific conditions. Information can be found at the Web sites for the program administrators. The program would apply only to LFGTE projects which are by definition cogeneration projects. Few LFGTE projects are cogeneration projects, but occasionally a cogeneration project could be configured.
- Innovative Peak Load Reduction Program: The program is administered by the CEC and can provide a $250/kW (net) grant to LFGTE projects. Information on this program can be found at the CEC Web site.
- Public Agency Loan Program: The program is administered by the CEC and provides 3% loans to municipalities, public agencies, and not-for-profit hospitals for renewable energy or cogeneration facilities. Information on this program can be found at the CEC Web site.

Greenhouse Gas Credits.   LFG consists of about 55% methane by volume. Methane is 21 times more potent as a GHG than CO_2 on a mass basis. The capture and combustion of LFG in a flare results in a significant net reduction in GHG emissions on a CO_2 equivalent basis.

The electric power–generation industry, and other major CO_2 producers in the United States, will have a difficult task in complying with international agreements intended to lower GHG emissions. Under certain conditions, LFG could be converted into transferable GHG-emission reduction credits. The cost of producing credits through LFG control could be less than the cost of many other methods of producing GHG-emission reduction credits.

If LFG is collected and flared, a GHG-emission reduction credit is produced by converting methane to CO_2. LFG flares are designed to achieve at least a 98% destruction of methane. Each ton of methane burned results in the net production of 21 tons of CO_2-emission reduction credits. If the methane is beneficially used to displace a fossil fuel, then additional emission reduction credits can be taken for each ton of LFG methane destroyed.

At the present time, a formal market for GHG does not exist in the United States; however, a few isolated GHG-emission reduction credit sales have occurred. These sales were reportedly in the range of $0.50 to $2.00/ton of CO_2 equivalent. Sale of GHG-emission reduction credits may provide a future source of revenue to support LFGTE projects.

Ownership and Financing of LFGTE Project Elements

Landfills.   Landfills are either publicly or privately owned. Public entities owning land-fills include cities, counties, authorities, or special purpose districts. Privately owned landfills are owned by one of the major national waste management firms or by firms which own one or a few landfills. All three of the major waste management firms—Waste Management, Allied, and Republic—own landfills in California.

LFG Collection and Control Systems.   LFG collection and control systems are usu-ally owned by the owner of the landfill. In some instances, particularly the older LFGTE projects, the developer of the LFGTE project installed and owns the LFG collection and control system. In such instances, the LFG rights lease called for the developer to install the well field necessary to support his or her project. In many of these cases, the landfill owner has installed a supplementary well field beside or around the developer's well field. The supplementary well field was required to fully satisfy the landfill owner's regulatory responsibilities. The well field supplying the LFGTE facility was not exten-sive enough to collect all of the LFG that could be recovered. On these landfills, there are two well fields—one owned by the owner of the LFGTE facility and one owned by the landfill owner. The LFG collected by the landfill owner's well field is flared.

It is also possible for a well field installed by the landfill owner to be under the temporary ownership of the owner of the LFGTE facility. If the well field predates the LFGTE facility, it is common for the well field to be leased or sold to the LFGTE project developer in the LFG rights agreement. This allows the LFG developer (or a third party) to become the gas producer and allows him to claim Section 29 tax credits.

Energy Conversion/Generation Systems.   LFGTE facilities can be owned by the landfill owner or by an independent developer. Since landfills are both publicly and privately owned, LFGTE facilities which are owned by a landfill owner can be either privately or publicly owned. Independent developers are always private sector firms.

For most private sector LFGTE facility power-generation projects, it is common to form a Gasco (gas-production company) and a Genco (power-production company). This arrangement is necessary to obtain Section 29 tax credits. The Gasco/Genco arrangement is not necessary if the private sector owner of the landfill retains the own-ership of the well field, sells LFG to the LFGTE facility developer, and claims the tax credits for himself. It is also not necessary to form a separate Gasco if the product of the LFGTE facility is medium-BTU or high-BTU gas, and if the product gas is sold to an unrelated party.

Financial Expectations of Investors

Municipal Financing.   If an LFGTE project is owned and operated by a municipal governmental body (e.g., city or county) or by a governmental agency (e.g., authority or special purpose district), then the cost of money for an LFGTE project is equal to the municipal body's cost of funds. The municipal cost of money is less than the cost of private money. The municipal cost of money has been in the range of 5% to 8% in recent years for 10- to 15-year loans.

The minimum expectation of a municipal body would be that the funds expended to finance a project would pay back the investment at the municipality's cost of money. In most cases, a municipality would expect to see a cost savings over this minimum return. The premium would provide a cushion against actual project performance below anticipated project performance.

A municipality would commonly conduct an analysis of project economic viability by selecting a minimum acceptable interest rate (e.g., 10%) and then run a traditional net present value or annual cost analysis to determine whether or not the project would yield a positive net present worth or a positive cumulative cash flow.

Private Sector Developers.   The cost of money for a privately financed project is greater than the cost of money for a municipally financed project. Corporations or individuals can invest equity (available cash) or can incur debt to finance a project. Debt can be backed by the full strength of the investor's balance sheet or the loan can be project specific or subsidiary specific with no collateral offered other than the assets of the project itself. The latter mechanism is known as *project finance, off balance sheet financing,* or a *non-recourse* loan.

Equity and balance sheet debt are generally valued equally by private investors since the amount of balance sheet debt which they can carry at any point in time is limited. A private investor must decide how to utilize his or her limited resources. Most private investors have investment alternatives not only in the LFGTE industry, but also investment opportunities in other industries. The lowest rates of return on investment are often demanded by utility subsidiaries, since utilities have historically seen low rates of return from their regulated businesses. Independent firms or firms which are owned by non-utility companies generally demand the higher rates of return seen in unregulated private industry.

Rates of return demanded by the private sector typically range from 15% to 25%. The expected rate of return is projected from a multi-year pro forma projection, which incorporates an internal rate of return calculation. The minimum rate of return that a private entity has at any point in time is often called its *hurdle rate*. If the projected rate of return exceeds the firm's hurdle rate, then the firm proceeds with project development.

Project finance involves a mix of equity and debt. A typical ratio of equity to debt is 20%/80%. The developer applies his hurdle rate only against the equity fraction. Project finance accomplishes two goals:

- It allows the private investor to leverage his limited equity (or equity equivalent) and undertake more projects.
- It makes less attractive projects more viable since the interest rate on a project finance loan is virtually always lower than a firm's hurdle rate. In effect, the margin from the debt fraction subsidizes the equity fraction, producing a higher rate of return on the equity.

The next subsection, titled "Independent Lenders," discusses the project finance concept in greater detail.

Independent Lenders.   Independent lenders will provide financing in the form of balance sheet debt or the form of project finance debt. Balance sheet debt is offered based on the overall financial strength and assets of a borrower, with varying degrees of consideration given to the financial viability of the project(s) that is (are) to be undertaken with the loan. For large, profitable companies with substantial assets, the lender may not be greatly concerned with the specifics of the project(s) to be financed. A lender's interest level in the details of the specific project(s) to be financed increases as the potential underperformance of the project(s) begins to have a significant impact on the borrower's ability to repay the loan. Many private LFGTE developers are highly leveraged or have limited financial support from their parent company, and often a true balance sheet loan cannot be obtained. A true balance sheet loan (e.g., one fully backed by a utility company parent of an LFGTE development subsidy) would probably be offered at 2% to 3% above prime rate.

Project finance (sometimes called *off-balance sheet* or *non-recourse* loans) requires extensive due diligence by the lender. Since the loan is guaranteed only by the project's performance and is collateralized only by the project's assets, the lender must satisfy himself that the project has no fatal flaws, and believe that the projections of financial performance made by the borrower are realistic. The projections are summarized on a pro forma balance sheet.

The pro forma balance sheets to support project finance loans show an equity and debt fraction. Equity is similar to the down payment on a mortgage. An equity fraction of 20% is common for project finance loans. The lender also wants to see that cash is available each year in excess of the amount required to service the debt. The excess is known as *coverage*. Lenders typically require that coverage exceed 20% of debt service during the life of the loan. The term of a project finance loan is generally in the range of 10 to 15 years. The interest rate for a project finance loan is about 4% to 5% over prime.

It should be noted that in addition to the above lender requirements, the borrower will evaluate the internal rate of return on the equity he or she invests. The minimum required internal rate of return on the equity fraction is typically in the 15% to 25% range.

A project finance transaction generally requires a great deal of legal, accounting, and technical support. The level of effort does not vary greatly with the size of the loan. Project finance is usually seen only on large projects or for a portfolio of smaller projects. Project finance can be facilitated by building the projects with equity or a balance sheet loan, and then converting to project finance after the project is operating.

Alternatives for Selling Power in California

As of July 2001, the date this report was prepared, the power sales market in California was in flux. California has taken measures, and will take measures, to solve problems that sprung from its deregulation of the electric industry. The subsections which follow discuss circumstances and opportunities as of July 2001.

Retail Deferral. The "best customer" for an LFGTE project is the host landfill itself. Maintenance shops, office buildings, scale houses, LFG blower/flare stations, groundwater and storm water pumps, and leachate treatment facilities are among the typical consumers of power. Other waste management facilities, such as transfer stations or recycling facilities, are sometimes located at a landfill. Occasionally, other potential retail customers (industrial facilities, office buildings, commercial structures, shopping centers, or public buildings) are located contiguous to or close to landfills.

The advantage of retail deferral, over other options for power sale, is that the base price is the utilities retail rate. Retail rates for large customers in California are in the range of $0.12 to $0.13/kWh. When power is self-generated, it is necessary to rely on the local utility for "standby" power. Power-generation facilities are not capable of operating 100% of the time. Downtime takes the form of planned and unplanned outages. As of June 2001, the CPUC was investigating standby rates as they pertain to distributed generation projects. The outcome of this proceeding is unknown; however, it is expected that standby charges for distributed generation projects will be reduced.

The cost of standby power typically reduces the effective net deferred cost by 10% to 20%. The specific impact must be calculated through application of the local utility's standby power rate schedule. Prior to the recent disruption of the deregulated market in California, retail customers leaving SCE and PG&E were required to pay a competition transition charge (CTC), which was scheduled to expire no later than 2003. The CTC was eliminated for San Diego Gas and Electric Company (SDG&E) customers in July 2000 since it was determined that SDG&E had recovered its "stranded costs." When SCE and PG&E began to pay more for power than they could recover from their rates, the CTC, because of the way it is calculated, ceased to exist. The future status of the CTC or other exit charges is currently unknown, and will not be known until all elements of the power industry–recovery package are in place.

The power requirements at a landfill are generally much lower than the power-generation potential at a landfill. Some of the power requirements at a landfill are intermittent (e.g., a recycling facility which only operates 8 to 10 hours/day).

If an independently owned facility is located on property immediately contiguous to the landfill, it is permitted to directly sell the facility power (with a direct power line to the facility). If a facility is not contiguous to the landfill, it is not possible to run a direct power line to the facility. An alternative to the direct power line is to construct a medium-BTU pipeline to the customer and to locate the LFGTE power plant at the customer's site. Power sales to off-site customers are discussed further in the subsection titled "Bilateral Contracts."

The limited revenue opportunity associated with on-site deferral can be handled in two ways:

- Construct a small project which is sized to satisfy on-site loads (such projects will generally be in the 30- to 300-kW range); or
- Build a large project and credit this 30 to 300 kW of consumption to the larger project. In all but the most stressed market conditions (such as those recently experienced in California), retail deferral has greater value than sale to the market.

Long-Term Contracts.   As of July 2001, California's investor-owned utilities were entering into power purchase agreements only for projects less than 100 kW in size. The contracts are as-delivered contracts and do not provide capacity payments. The rate paid for the power varies monthly and represents the utility's short run avoid cost (SRAC). The SRAC is directly pegged to natural gas cost. These contracts are continuously available and offer a convenient way for a very small LFGTE project to sell excess power. The contracts do not have the comfort of offering a fixed price.

Contracts for larger projects are only being offered by California's Department of Water Resources (DWR). As a result of the questionable creditworthiness of the California IOUs, the DWR has stepped in as the short-term and long-term buyer of power. It has been reported that the DWR has been signing 10-year contracts at an average price of $0.07/kWh; however, the details of the agreements are considered to be confidential.

The DWR has been concentrating on securing large blocks of power, and to date has had no preference for "green" power over "brown" power. Projects less than 10 MW in size do not appear to be large enough to be of interest to the DWR as of June 2001. If the DWR is installed as the long-term purchaser of power, the DWR might develop standard-offer-type contracts for small generators. In the interim, it appears as if only large projects or developers who are able to aggregate several projects could obtain a contract with the DWR.

The creditworthiness of public utilities (e.g., Los Angeles Department of Water and Power [LADWP] and Sacramento Municipal Utility District [SMUD]) and public power agencies has not been called into question and these entities continue to purchase their own power. The procurement methodology at public utilities varies, and in some cases it may only be possible to obtain a contract through a formal request for proposal/bid process. LADWP has specific interest in green power, and they are willing to pay a premium for green power. Additional information on green power sales can be found in the subsection titled "Green Power Sales Opportunities."

Sale to the Spot Market.   With the termination of the state-related power exchange (PX) in February 2001, a great deal of fluidity left the California power market. The power market as of June 2001 largely consisted of the Automated Power Exchange (APX) and the state-related Independent System Operator (ISO) shortfall purchases. If an owner of an LFGTE project intends to sell power to the open market, the following steps would need to be taken:

- Register with the ISO as a participating generator. The generator must be responsible for at least 1 MW of gross capacity.
- Install an ISO-approved meter at the facility (or facilities).
- Obtain a power-marketing license from the Federal Energy Regulatory Commission (FERC).
- Sign an agreement with a power market (such as the APX).

Bilateral Contracts. A bilateral contract is a power sales contract between a generator and a retail customer. Power sale to a customer located adjacent to a landfill or at an off-site location through a medium-BTU pipeline project would be covered by a bilateral agreement. A second form of a bilateral agreement would be one between a generator and a customer not directly connected to the generator. The power which is sold would travel through the utility grid. In order to consummate such a transaction, even when only one retail customer is involved, the generator would be required to register as an energy services provider with the CEC, register with the ISO, and install an ISO-approved meter. As of June 2001, the elimination of direct retail sales was being discussed as a component of the effort to restructure the California utility industry.

Green-Power Sales Opportunities

Direct Subsidies.   Some states, particularly those which have or will undergo deregulation of their power industry, can or may provide transitional assistance to the renewable power industry. California is such a state. The California direct subsidy program was previously discussed.

Market-Based Support.   There is some market-based support for renewable energy on the retail level in the form of customers willing to pay a premium for green power. Some cities, companies, or government agencies have made commitments to buy green power, and some residential customers are attracted to green power. Green power reaches customers in two ways: (1) power marketers in states with a deregulated power industry often sell renewable power in addition to brown power to retail customers; and (2) some utilities in regulated markets offer their retail customers a renewable power purchase option. As of late 2000, green power was being marketed by retailers in six states (California, Pennsylvania, Maine, New Jersey, Massachusetts, and Connecticut). In early 2001, in a notable reversal for green power, major green power retailers in California abandoned the retail market and returned their customers to their respective IOUs. At the present time, only selected municipal suppliers, such as LADWP, offer a green-power option to their retail customers.

Green power on California's APX traded at a premium of only about $0.003/kWh above brown power over the last 2 years. It will probably take several years before green power premiums make much of an impact on the LFGTE industry. Green-power demand must first outstrip green-power supply. In the current California market, LFGTE does not experience a direct financial benefit simply because it is green power. It is expected that this will change over time.

Current Green-Power Sales Opportunities.   As discussed above, revenue enhancement through green-power sale offers limited potential at the present time; however, there are currently three ways for LFGTE projects to sell green power:

- Respond to utility green-power requests for proposals (RFPs).
- Enter into contracts with companies who retail green power.

- Execute bilateral contracts with large power users who have a commitment to green power.

The LADWP recently issued an RFP for the procurement of green power. Price is a major consideration in this and other RFPs. A wealth of up-to-date information on green power marketing can be found at http://www.eere.energy.gov.

Chapter 12

Competition for the Microturbine Industry

With respect to competition, the "why" questions are at least as important as the "what" and "which" questions. "Why" questions ask, for instance:

- Why might a large corporation buying a new technology firm *not* be a positive sign for that technology's growth?
- What corporate patterns do consumer/investors watch for, among large oil, power, and manufacturer companies, to know whether they can expect continued parts and service for their new technology equipment?
- Can government funding make up for large corporations' relative lack of support of distributed technologies?
- What new distribution technologies are better supported than others?
- What is the tax credit mix?
- For how long are the tax credits in effect?

…and so on.

So, this chapter could also be entitled "Which new distributed energy source will get the largest share of investment and business?"

For subjects that relate to distributed energy methods, risk, investment, competition, technology trends, and technology track records all are inseparable subjects. This is because the field is relatively new. If the subject were conventional gas turbines or steam turbines, these subjects would all have developed their own staff and infrastructure within manufacturing companies and end-user firms. New technology personnel have to be more versatile and much quicker on their feet.

That distributed generation has a great long-term future is undeniable. That certainty stems from the needs of both the world's poor and its affluent. The reasons for that paradox are varied.

About two billion people globally live without electricity because their home areas have no transmission lines. These people are often without resources desired by the western world, such as oil, gas, and mineral wealth, so the megacorporations have not yet raced to their shores. Or in the case of people in China and India who are not yet part of their "new middle class" courtesy of a relocated western sector job, the corporations have not yet expanded that far.

Today, there are many proven successful or at least highly promising, new distributed generation (small stand-alone grid-not-required generation packages) commercially available. This means that transmission lines in these power-impoverished regions may never get built. Unused to having any power, newly emerging communities may be glad to have a new hospital that runs on a microturbine or a new factory that runs on power from its wind turbine, even if those power sources are not always as reliable as the power generated by the power grid in the United States.

Some U.S. areas, however, say the word *reliable,* when used to describe their power grid, is at best a euphemism. Californians, with recent rashes of crippling brownouts and squabbles with Texas executives who gouged them on power tariffs, are a case in point. So even the officially power-rich now want their own, albeit smaller, power-generation source for additional security. In one case, mentioned in Chapter 9, Microturbines Operating "Solo," the power customer was a Japanese electronics factory that opted to have two distributed energy sources that they owned and controlled as primary power and first backup source. The power grid was their *second* backup.

According to "North American Renewable Energy Markets" (Frost & Sullivan), the region's revenue for renewable energy sources ($204 million in 1998) was $843.3 million in 1999. Most of this was attributable to wind power projects, favored by a U.S. federal production tax credit.

So the needs of the industrialized wealthy are likely to drive alterative energies' growth as much or more than the developing world. There are also other factors, the environment and global warming being a major one. Since governments understand that altruism is not endemic in capitalistic societies, they impose laws or provide incentives that produce the same result. These include but are not limited to the following:

- Emissions taxes for nitric oxide (NO_x), sulfur oxide (SO_x), and carbon dioxide (CO_2) per unit weight of emissions
- Grants, and "good boy" certificates for energy conservation (see Chapter 10, Combined Heat and Power With Microturbines)
- Loans and incentive programs
- International/national bodies, associations, and protocols, such as Kyoto, Rio, and Montreal
- Tax credits for being a small power producer
- "Buy from" tariffs for small power producers who make a power excess that they can sell back to the grid
- Deregulation and privatization initiatives

Governments in countries such as the United States may also develop an emissions trading scheme. This legislation may or may not belong on the same list as the above items, depending on whether it actually reduces a country's contribution to the global warming load or promotes any of the above-mentioned measures.

As what used to be the "great world out there" shrinks to the "global village," corporations are more internationally image conscious. As evidence: Texaco CEO Peter Bijur, in June 2000, said that climate change is "without question, the single greatest environmental challenge we face.... This, then, is the emerging profile of our industry— one that will harness the profit motive in the service of the environment." This could be good news too late for the Ecuadorians in their Amazon rain forest home, who still struggle with badly managed oil residue that has polluted everything they know, including their drinking water.

In 2000, Texaco—for $67 million—acquired a 20% equity interest in Energy Conversion Devices (which dealt with energy storage and hydrogen technology). The resultant joint venture, Texaco Ovonic Fuel Cell, deals with fuel cells. BP-Amoco, also keen on its environmental image, has invested over $150 million in solar photovoltaics. Royal Dutch/Shell had spent over $500 million total on renewable energies, as of 2002.

These expenditures need to be studied carefully to determine what the lasting effects for small consumers will be. The money to buy a small company that has developed an alternative technology to "oil and gas guzzling" in some way is a drop in the bucket for an oil giant.

Economic conditions in the last quarter of 2006 included (but were not limited to) the following:

- War in Iraq (even though the United States has been getting what other countries have called "more than their fair share" of Iraq's oil)
- Threatened war in Iran
- A still-in-construction pipeline from Azerbaijan (via Georgia and Turkey)
- Rich fields off Angola not yet in full production

These conditions have produced gasoline prices in excess of $3 a U.S. gallon in the United States. Europe was already used to higher prices, but its prices (between $4 and $6 for a U.S. gallon, currently) are rising, too.

If one were to put oneself in the position of an oil company executive whose bonus may depend on annual profits, one may find that one is inclined to insulate the alternative energy technology that one just bought from major market success. If, for instance, photovoltaic cells, wind turbines, or even geothermal heat pumps were to cut per capita use of fossil fuels by 50%, it might spell disaster, at least on the horizon, if not imminently, for an oil company executive. It is a different story if the executive is adaptable enough to take a position as CEO of a solar cell company, but given the economies of scale, this represents a considerable step down.

This is particularly true in the United States, where a CEO may make over 110 times what his average worker makes. In Japan or Sweden, where the figure is more like 11 to 13 times as much, the executive may move without a backward glance. Besides, in those cultures, if for different reasons, more civic-minded behavior that favors the poor man is the norm, their "autopilot" mode.

In other words, several of these innovative alternative technologies were close to market ready, if not already there, when large oil bought them. In today's market, big oil's reaction to their newly acquired subsidiaries may be "give them a research budget and ask them to 'perfect' some more." The average U.S. consumer needs to understand these alternative energies to squawk in meaningful terms, that he or she is being overcharged for anything. In a country where the illiteracy rate is over 35% and climbing, the "public voice with government" is left to a shrinking body. The typical U.S. consumer now finds him- or herself in a quickly changing world, where electoral speeches promising "new energy" in some "future" have to be enough, because he/she cannot adequately debate the subject.

Nothing can stem the advance of alternative technologies, struggling and fraught with development problems though they be. The rest of the world outside the United States, developed or not, will see to that. They have to, as they do not have the military might to ensure their supply of conventional fossil fuels.

The same logic that applies to oil giants also applies to "large power" giants. In fact, they are increasingly the same giant. Shell is now a major player in the U.K. independent-power-producer (IPP) scene and its own best customer for its fossil fuel. The logic extends further to giant, primarily turbomachinery-fostered giants, such as General Electric (GE). Economies of scale dictate that the after-sales spares market, which totally eclipses actual turbine sales, is far greater for large turbines than solar cells or microturbines. Furthermore, although GE is not a construction or fossil fuel company, it has joint ventured with all companies of these types and sells them insurance when it can. Also, the financial arm of GE is everywhere, for all industries, large and small. "Small" business is pursued only when the larger margins afforded by "big" business patronage are already dealt with or sheer numbers make "small" worth pursuing.

GE, for instance, has investments in microturbines. Via planned acquisition of Honeywell, GE's smallest power-generation unit was to have been 75 kW for a while. Then

Honeywell got out of the microturbine business and began buying back all of its microturbines. As of early 2007, their Web site does not reference a microturbines service facility.

In other words, the strategy of very large corporations may include acquisitions of new technology companies. That then buys control of that technology's growth. Those Cinderella new technology divisions can then stay on life support until major tax credits or funding of some sort make it appropriate to send the Cinderella(s) to the ball.

So far, Asea Brown Boveri (ABB) is the only global giant to leave the business of turbomachinery. That turbomachinery business left with the portion of ABB that, with Alstom, became Alstom Power in 1999. (A year later, ABB sold out to Alstom for about $1.5 billion.) ABB, however, still deals "on their own" with support control systems for conventional power plants and accessory machinery, such as fans. ABB now also belongs in renewables technologies, efficient distributed generation from fossil fuels, and related transmission systems.

ABB forecasts that revenues in distributed power will reach about $1 billion in 2005 and $2.5 billion in 2010. The money is significantly less than in the power-generation business that ABB offloaded, but so is initial investment in distributed technologies.

According to a Wall Street investment company: "… the power industry is *now* where the telecommunications industry was in 1984 …. Energy Technology is a long-term play."[*] The next 20 years are likely to see roughly $10 trillion worth of energy infrastructure investment (Figure 12-1). Some estimate the market opportunities for distributed energy could reach around $200 billion a year.

So stock prices of distributed energy companies can and do fluctuate wildly. Capstone, CA, one of the better established microturbine players, has had times when its stock price has dropped.

Where do microturbines stand in this melee? Broadly speaking, they stand as follows:

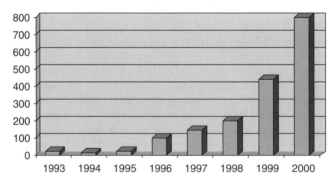

FIGURE 12-1   *Venture capital investment in distributed energy ($ millions). (Source: Capital E. 2001. A distributed energy primer. Available at: http://www.cap-e.com/Report/default.cfm.)*

---

[*] Quote: Capital E. 2001. *A distributed energy primer.* Available at: http://www.cap-e.com/Report/default.cfm.

1. Initial "prototype" nature problems gave the microturbine an initial "cute but you can't rely on them" rap. This is normal for any new machinery type or substantially different model.
2. Microturbines often need to use fuel that is "waste" or somehow "free" to justify their own purchase. Irregularities in that fuel supply can cause problems and downtime (see Case 11-1 in Chapter 11, Unconventional Microturbine Fuels).
3. General knowledge among the average U.S. citizen does not extend to an understanding of what "CHP" means. You may be sure it does in Europe or Japan. This unfamiliarity limits market growth the United States. Most people are dimly aware of what the Department of Energy is, but not to the point that their Web site is stored as one of their "Favorites."
4. Microturbine emissions rates on a parts-per-million (ppm) volumetric basis have been good and on the order of 9 ppm.
5. The microturbine's size makes it ideal for direct waste heat applications such as greenhouse hot air supply.
6. Grid interconnection barriers, including utility rules that attempt to redefine reason, are a barrier to market growth.
7. Many microturbines have proved to be reliable, to an extent that mitigates item 1 in this list.
8. Despite microturbines' progress, other alternative energies have made as much or more and may show more promise.

COMPETING DISTRIBUTED ENERGY SYSTEMS

Reciprocating Engines

One of ABB's generation contracts was to install 10 combined heat and power (CHP) units in the United Kingdom for Scottish and Southern Energy plc. The $100 million deal specifies ABB as the operator of the plants for the first 17 years. The installations, which are to use natural gas–fueled reciprocating engines, will have capacities up to 750 kW each.

Some of the units will be placed in greenhouses, so as an added touch, those turbines will recycle exhaust CO_2 to stimulate plant growth.

So if the application permits their size, reciprocating engines are a viable alternative to microturbines. Given their respective fleet maturities, reciprocating engines may be the "familiar comfort" choice.

Fuel Cells

There has been a great deal of hype and consequential high investment in fuel cells. That said, companies such as Plug Power, a fuel cell manufacturer, have seen their stock price fluctuate to the same extent that Capstone's has. When everyone gets the problems ironed out, it is true that pure hydrogen makes a very clean fuel, and if it is cheap and universally available, it will make fossil fuels exit faster than one can say "dinosaur." Its problems, however, are achieving *pure*, *cheap*, and *universally available*.

Right now, hydrogen can manage *pure* but not *cheap*. Fuels that provide a hydrogen source may also contain sulfur impurities. That in turn causes hydrogen sulfide (H_2S) that then combines with moisture to form sulfuric acid. That makes all kinds of messes with hardware it touches (see Chapter 11, Unconventional Microturbine Fuels). Removing sulfur and other impurities are some of the issues that get in the way of *cheap*. Even if it were cheap, it may take years to be universally available.

Earlier this year, information on fuel cell progress to date was provided as follows.[*] There are four main types of fuel cell with corresponding characteristics:

- Polymer electrolyte membrane
  - Size: 1 to 10 kW
  - Efficiency, 25% to 40%
  - $5000/kW cost
  - Applications being developed for telecommunications and battery replacement
- Phosphoric acid fuel cell
  - Size: 200 to 1000 kW
  - 40% efficient
  - $3500/kW
  - Tests indicate reliability/availability are good, but high cost is a drawback
- Molten carbonate fuel cell
  - Size: 250 to 1000 kW
  - 48% efficient
  - $3000 to $4000/kW
  - Trial system stage
- Solid oxide fuel cell (SOFC)
  - Size: 1 to 250 kW
  - 48% efficiency
  - $10,000 to $20,000/kW
- Trials done and continuing

In places where grid power is unavailable or very expensive and/or gas/fuel oil is expensive, a smaller SOFC can pay for itself in about 2 years.[†]

From an SOFC in a remote location, a 5-kW power plant is being developed with oil at $2/liter and no power available. The time to return on investment is 2 years for a $20,000 cost.

For an SOFC with power at $0.11/kilowatt-hour (kWh) and gas at $7 per million cubic feet, payback time is 2.2 years.

Hybrids

A *hybrid* is a combination of a microturbine and a fuel cell. When these are in combination, the microturbine's 25% to 40% efficiency jumps to about 80%.

Not only can hybrids be used "stationary" in power-generation applications, they can also drive cars, buses, trains, and earth-moving equipment. They have their drawbacks, which include the initial high price tag. As fleet size and familiarity grow, so will the price tag.

Photovoltaic Systems

Several companies in different countries now sell components to put together solar-power generators that can fit on the roof of a house or a small business. They can be made larger or smaller by addition of standard panels. The return on investment for a house- or small business–sized system in California was about 7 years, allowing for

---

[*] Source: Electric Power and Research Institute. 2005. Status, trends and market forecast for stationary fuel cell power systems. Ohio Fuel Cell Symposium.

[†] Source: Fuel Cell Technologies Corporation. 2002. *Reliable and affordable remote energy conference*. Fuel Cell Technologies, Ltd. Kingston, Ontario, Canada.

a federal tax rebate (available for businesses) and state rebate, given end-of-2004 fuel (gasoline) prices (roughly $2.25/U.S. gallon). Given that average price rose to $3.40/U.S. gallon and higher in subsequent years, many consumers may give this energy source another look.

Most of the component manufacturers for solar-power systems manage their economies of scale by offering courses at about $400 per person. The course then gives attendees a license to sell the systems, enough knowledge to assemble them, and a nearby parts supplier.

Complications may arise from requirements that these rooftop systems must, by law, not look ugly to neighbors.

Turboexpanders[*]

The basic difference between turboexpanders and steam turbines is that turboexpanders contain metallurgy that lets the gas in question be highly corrosive (an acid gas, for instance) or otherwise unsuited for a conventional steam turbine. This is not as well-known a field as the turboexpander's capabilities justify. That may have to do with the public relations capabilities of its pioneers. Nevertheless, their technical abilities means that there are now competent manufacturers who can custom make a one-of-a-kind turboexpander that can cover a power-generation (or mechanical drive) range of about 3 kW to over 1 megawatt (MW) (a few horsepower to well over 1000 horsepower).

All one needs (besides the turboexpander system) is a gas that may already be pressurized (such as geothermal steam). Any gas that metallurgy can handle will work, as will inlet pressures of as low as 100 psi and over 3000 psi. Gas temperatures can be from 600°F to a few degrees above zero.

The turboexpander therefore does not need any combustion section or fuel supply, which limits the number of parts exposed to flame temperatures.

Ultimately, the choice is made by a series of factors including but not limited to the following:

- Familiarity with manufacturers and applications
- Availability of a source gas stream at "excess" pressure
- Price
- Service (The track record of many microturbine applications has been excellent; see Bloch, H. and Soares, C. 2001. *Turboexpanders and process applications*. Gulf Publishing. Woburn, MA.)

Wind

According to the American Wind Energy Association, more than 3600 MW of wind power capacity were added globally in 1999. This is a 36% increase over the previous year.

Innovations in wind technology continue. ABB, for instance, has designed the Windformer. The system reduces maintenance and equipment costs by eliminating both the gearbox and the transformer. Preliminary claims include an electricity cost of less than $0.04/kWh. That makes it competitive with conventional gas-, coal-, and oil-fired power plants. No wind turbine has done that yet.

---

[*] Source: Bloch, H. and Soares, C. 2001. *Turboexpanders and process applications*. Gulf Publishing. Woburn, MA.

ABB is testing a 500-kW prototype. Vattenfall, Stockholm, was assigned to install a 3-MW demonstration Windformer. ABB plans to avoid problems with changing frequencies of current (variable winds) by converting the power to high-voltage direct current (HVDC) for transmission by using HVDC Light.

HVDC Light uses semiconductor technology and controls to eliminate the problem of incompatible frequencies by doing away with frequency altogether. HVDC can cut transmission losses by 50%, but generally has been practical only on a large scale, ABB said. With ABB's large-scale DC power transmission link between a substation in northern Argentina and another in southern Brazil, because the 1000-MW link is DC, the two electric systems' different frequencies do not matter.

ABB also received a $120 million order to lay a DC transmission line under Long Island Sound to link the power grids of Connecticut and New York. The connection will use HVDC Light technology to carry 330 MW over 40 km. So ABB's logical claim is that its HVDC Light system can economically transmit the power output of a small generator, perhaps a wind farm, as far as 40 km.

ABB also states that the Internet makes it possible to monitor and maintain microgrids, such as wind farms, remotely. That this is true was comically reinforced in 2002 by technicians from GE Wind who, perhaps accidentally, got the data from their internal Intranet (which controlled their wind farms) onto the Internet.

Remote control of any fossil-fuelled turbine is not a new subject and dates back a few decades. This is more a case of Don Quixote's old sparring partner catching up with conventional turbine technology when it is a large enough unit or part of a wind farm. Simple house-sized wind turbines do not yet have sophisticated controls, but that is in the cards, too.

WIND HYBRIDS AND OTHER SOURCES

Some extracts of an excellent study done by the government of the state of Iowa[*] follow. The study pivots on that state's wind and hydro resources, but as Iowa is a farming state, it also includes work on biodiesel. Because this is a potential microturbine fuel, that extract is also included.

For convenience, the study's cost data are included in Chapter 13.

Wind Hybrid Technology Definition

A wind hybrid energy system incorporates wind energy together with one or more forms of electric generation or energy storage to serve a specific energy load. A wind hybrid system revolves around a wind turbine with other energy sources or storage to help deliver energy on demand. For example, a diesel generator may be controlled to operate when the wind turbine is not producing sufficient power to meet the load requirements. Or a battery bank may be connected to the turbine to store excess power being generated. A combination of technologies can be used together, such as solar panels, batteries, a generator, and a wind turbine. Also, there are some interesting emerging technologies, such as flywheels and fuel cells, which promise to help make better use of intermittent low-cost wind energy.

---

[*] Source: Iowa Wind Energy Institute. 2001. *Wind hybrid electricity applications*. Iowa Department of Natural Resources. Des Moines, IA.

In this section all these options will be examined, from the utility scale to the residential scale. The Iowa Department of Natural Resources (IDNR) looked at some applications where different forms of distributed generation are blended into a utility's energy portfolio to complement wind generation. Specifically, IDNR examined re-powering decommissioned dams to create wind-hydro solutions and the blending of wind with the output from diesel peaking generators owned by many utilities. Then a review of commercial and home-scale renewable energy and energy-storage technologies was undertaken. IDNR concludes with a cost-benefit comparison of all technologies to assist the reader in better understanding the applications and economics of different forms of energy generation and energy storage that can be coupled with wind generation. In many cases, the economics may not be favorable at the present time but may become favorable in the evolving environment of electric deregulation and changing fuel costs.

Advantages and Disadvantages of a Wind-Hydro Hybrid

The potential advantages of combining the two renewable resources wind and hydro in a project include the following:

1. The hydroelectric generation is much more stable from minute to minute, hour to hour, and day to day.
2. The combination would provide a high percentage of renewable power to the owner's electric load.

The disadvantages of combining these two resources include the following:

1. Since hydro generators are on rivers that usually are at lower levels, the location may be some distance from higher elevations more suitable for installing a wind turbine. A distribution line would be needed to connect the two facilities to the customer's facilities.
2. The combined output of the hydro and wind generators needs to be about the same or less than the customer needs, because any excess power sold back to the utility will not have much value due to the relatively low avoided cost buy-back rates.
3. The grid will still be needed to provide a steady source of power, unless the facility uses an engine-driven generator so that it can operate completely isolated from the grid.
4. Since the hydro capacity is fixed by the height of the existing dam and the river flow, it cannot be expanded to meet any increased needs of the customer.

Case Study

The recent re-powering of the Mitchell dam provides some cost data and a basis for a case study for a wind-hybrid combination. The sketch in Figure 12-2 illustrates this case study.

In this study, it is assumed that a manufacturing facility is located within a half mile of the Mitchell dam. Furthermore, the facility has enough property with good wind exposure to install a Vestas V44/660-kW wind turbine beside its facility. If this manufacturer placed a high premium value of using renewable energy in its manufacturing process, then it could re-power the dam and install the wind turbine to help

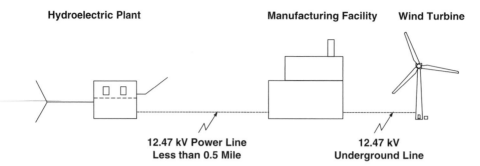

FIGURE 12-2   *The Mitchell dam: a wind-hybrid combination. (Source: Iowa Wind Energy Institute. 2001. Wind hybrid electricity applications. Iowa Department of Natural Resources. Des Moines, IA.)*

supply the energy needs of its facility. The Mitchell dam will provide an average of 2.9 million kWh and the wind turbine would average 1.7 million kWh. Together they net 4.6 million kWh.

For this combination to be utilized well, the manufacturing facility would need to use a third to half more energy than this, assuming the facility does not receive net billing from the utility. A facility this size might be the largest electric customer in this rural county. The facility would be on a two-part rate with both demand and energy charges and would likely be paying a relatively low rate for electricity due to its size. Assuming the customer retained electric service from the utility, the wind-hydro generation would reduce the facility's energy charges each month and would normally reduce the customer's demand charges to some extent, depending upon several factors. In this case study, it is assumed that the customer pays an average of $0.05/kWh under a two-part rate, with the energy component at $0.03/kWh and the demand component at $9.00/kW-month. Table 12-1 illustrates the power used, generated, and purchased with and without the wind-hydro combination, based on the several assumptions and estimates.

Based on several assumptions, the manufacturing facility would save about $200,000 per year in power costs and income taxes due to the federal PTC. The actual cost of the Mitchell hydro facility was $813,000 and the cost of the wind turbine would be about $800,000 (Table 12-2). Additional electrical lines and interconnections would bring the total capital cost to about $1,650,000. The estimated operating cost of the hydro and wind generators is $75,000. This would include labor, parts, equipment reserve, insurance, property taxes, and lease of the dam site. If the $1.65 million investment were financed with a 20-year loan at 9% interest, the annual loan payment would be $181,000, which is more than the operating savings. The result would be a negative $56,000 cash flow during the loan payback period.

If the income tax impacts are considered, then the quick 5-year accelerated depreciation of the wind turbine provides significant after-tax savings for 5 years, assuming the facility's owners have enough income tax liability to take advantage of this depreciation. The operation cost savings would of course be larger if there was some premium paid for carbon emission reductions or if the interest rates were lower.

In summary, the wind-hydro project would result in a pre-tax loss, but an after-tax savings for most corporate owners. However, the calculations were based on a 20-year loan payback. Most industrial facilities require their investments to pay off in a matter

TABLE 12-1   Power Used, Generated, and Purchased With and Without the Wind-Hydro Generation

Energy	Prior to Wind-Hydro	With Wind-Hydro
Annual kWh Needed	7.0 million kWh	7.0 million kWh
Annual kWh Purchased	7.0 million kWh	3.4 million kWh
Annual kWh Generated	0.0 million kWh	4.6 million kWh
Annual kWh Sold to Utility	0.0 million kWh	1.0 million kWh
Cost of Purchased Energy	$210,000	$102,000
Demand		
Average Monthly Demand	1300 kW	1300 kW
Average Demand Supplied by Utility	1300 kW	775 kW
Average Demand Supplied Wind-Hydro	0 kW	525 kW
Cost of Purchased Demand	$140,400	$83,700
Total Cost of Purchased Power	$350,400	$185,700
Revenue from Power Sales to Utility	$0	$20,000
Production Income Tax Credit	$0	$17,000
Net Savings in Power Bill and Taxes		$202,000

Note: There may be corporate tax benefits from the 5-year accelerated depreciation of wind turbine.
(Source: Iowa Wind Energy Institute. 2001. *Wind hybrid electricity applications.* Iowa Department of Natural Resources. Des Moines, IA.)

of a few years at most, not 20. If the owners placed a high value on using renewable resources, then this slower payback on electric generation infrastructure may be an acceptable premium. Although Iowa does not have much hydroelectric resource potential, a few of the better sites will likely be developed if the market price for this type of renewable resource power increases to the $0.05 to $0.06 range. Because of the steadier output that can be controlled to some extent, hydroelectric power will always have a higher value to the electric system than wind-generated power.

TABLE 12-2   Mitchell Hydro Project Cost

Cost of hydro facility	$813,000
Cost of wind turbine	$800,000
Cost of interconnection equipment	$37,000
Total loan amount	$1,650,000
Annual Profit/Loss	
Net savings on power purchases	$200,000
Operating costs	($75,000)
Loan payments (20 years at 9%)	($181,000)
Net income/loss per year	($56,000)

(Source: Iowa Wind Energy Institute. 2001. *Wind hybrid electricity applications.* Iowa Department of Natural Resources. Des Moines, IA.)

Utility-Scale Wind-Diesel Hybrid Systems

Introduction and Background

For many years, wind generators have been connected to small isolated electric systems around the world that are powered by diesel generators. There are numerous technical reports about how various types of wind-diesel systems can be operated to provide a steady continuous source of power. In nearly all but the most remote rural locations in the U.S. mainland, the electric grid is available to supply power where needed. Therefore, a diesel engine generator would usually not be justified unless it was for backup emergency power. There are a surprising number of diesel backup generators installed for that very purpose in the U.S. and the number continues to increase as the reliability of electric service becomes more essential to various businesses, government services, and institutions. In some cases, utility incentives help pay for some part of these backup generators when they can be brought online to help reduce the utility's peak load.

Large 1- to 4-MW diesel generators now cost about $350 to $400/kW installed. Many are being installed now primarily by smaller consumer-owned utilities to meet local peak load demands. Most are operated on #2 diesel fuel oil and run for 50 to 200 hours/year, depending upon peak load. With #2 oil costing $0.75 per gallon, the operating cost is 5.5¢ per kWh. The fixed ownership costs are about $50 to $65/kW-year for the 1- to 4-MW–sized units, depending upon the capital amortization period.

If grid power is available, the high operating costs of engine-driven generators relegate their use for peak shaving and emergency conditions. Occasionally gas-fired engine driven generators are used to convert landfill or biogas to electricity on a baseload basis.

Biodiesel fuel is a blend of processed soybean oil and diesel fuel. Blends are typically from 20% soy oil (B20) to 100% soy oil (B100). If a diesel generator were run on 100% soy oil, then it would be essentially 100% powered by a renewable fuel, other than the small fraction of crankcase lube oil that is consumed by combustion. A 100% soy oil fuel costs about $2.40/gallon for tanker loads. Since soy oil is essentially a by-product of soybean processing, its price is not as sensitive to the cost of soybeans. Cost is more sensitive to supply and demand for soy oil. Based on a longer-term estimate of $2.50/gallon, the operating cost of the diesel generator on 100% soy oil would be about $0.17/kWh, instead of $0.055 with diesel fuel.

Potential Applications of Wind-Diesel Hybrid Systems

Since the grid power is available everywhere in Iowa, an engine-driven generator would only be used during peak load periods or emergencies. One potential application with a wind turbine would be to use an engine-driven generator to manage a facility's peak demand charges from the utility. The wind turbine would provide power to reduce the facility's energy purchases. If the wind turbine were not generating during a peak demand period, then the engine-driven generator would be automatically started. This application would only be cost effective when the utility's tariff had a specified time period, such as a 4-hour window from 3 PM to 7 PM, where the demand was measured. A computer could then determine if the engine-driven generator would be needed during that time period. The engine-driven generator run time would be limited to a few hours per day if the wind turbine was not generating and the facility was at a high load. Many utility tariffs do not have a specified time window where the demand is measured. Under these tariffs, the facility could set their peak demand at any hour of the day.

If the facility does not pay a demand charge, but only an energy charge, then there would be no reason to manage the demand or add an engine-driven generator. In general, if the utility's tariff has no demand charge and only an energy charge, then there is more incentive to install a wind turbine. If the utility has a high demand charge and low energy charge, then adding a wind turbine is probably not economical. Wind turbines are good energy producers, but poor demand reducers.

The Lac Qui Parle School in Minnesota has a wind turbine and engine-driven generator. A 225-kW wind turbine was added along with an 275-kW engine-driven generator to manage the school's peak demand, which was measured during a specified period of time. When the net load of the school less the wind turbine's output was above a target demand level during the demand measurement period, the engine-driven generator was run to reduce the demand below the target level. Due to the lack of significant financial incentives, this demand reduction scheme is not currently being used. Figure 12-3 shows the school and wind turbine, and Figure 12-4 shows a 60-kW diesel generator set.

Biodiesel fuel could be burned in these types of installations, but the school or other facility would still consume a lot of energy from the utility that would not be generated by renewable energy. Since the biodiesel fuel costs significantly more than regular diesel fuel, the system should be designed to minimize the engine-driven generator run time.

Advantages and Disadvantages of Wind-Diesel Hybrid Systems

The advantages of adding an engine-driven generator to supplement a wind turbine is the ability to reduce a facility's demand charge from the utility. If the utility's tariff is designed to encourage this management of demand, then a wind-diesel system might be economical. The wind turbine would be responsible for reducing the energy purchased, while the diesel would reduce the billing demand charge. The main disadvantage of this type of hybrid system is the high operating cost of the engine-driven generator. Any significant amount of run time will likely make the additional installation of the engine-driven generator uneconomical.

Case Study of Wind-Diesel Hybrid Systems

The following case study shows a school system's purchase of power under a two-part rate with a high $15.00/kW demand charge and a $0.03/kWh energy charge. The average

FIGURE 12-3 *Lac Qui Parle School wind turbine. (Source: Iowa Wind Energy Institute. 2001. Wind hybrid electricity applications. Iowa Department of Natural Resources. Des Moines, IA.)*

FIGURE 12-4 *60-kW diesel generator set. (Source: Iowa Wind Energy Institute. 2001.* Wind hybrid electricity applications. *Iowa Department of Natural Resources. Des Moines, IA.)*

cost of energy was $0.076/kWh. The utility only measures the school's demand from 4 PM to 8 PM Monday through Friday. The school normally has a 250-kW monthly peak demand, except during the summer, and it uses about 1000 MWh/year. Their electricity cost is $75,000/year. The school has decided to install a new 250-kW wind turbine costing $375,000. It is projected to generate 613,000 kWh at this location. Based on a 15-year loan at 6% interest, the loan payments and operating expenses of the wind turbine total $47,000/year.

As Table 12-3 indicates, adding an engine-driven generator backup in this particular instance does improve the economics. However, installing a new 250-kW wind turbine increased the total cost of operations. In order for a wind turbine and engine-driven generator to both be economical, high demand charges as well as high energy charges would be needed. A utility incentive program for customers to install backup generators that can be used to reduce the utility's peak demand would also improve the economics. If a utility has a high demand charge and a low energy charge, then it will probably not be economical to add a wind turbine, since a wind turbine primarily reduces energy purchases.

Observations and Conclusions on Wind-Diesel Hybrid Systems

Diesel generators vary widely in cost with different sizes. A typical residential-scale backup generator might be a 3-kW-size unit costing about $1500. These units are not meant for continuous use and are more available to run on gasoline. A commercial-scale unit installed and ready for continuous operation would cost closer to $30,000 for a 25-kW unit. On the utility scale, a 2000-kW unit would cost about $700,000, which yields a low per-kW price of $350.

There will likely be few instances where the addition of an engine-driven generator will improve the overall economics of a wind turbine installation for an electric customer in Iowa. Wind turbines are most economical where the energy charges are high. Since wind turbines produce intermittent power, they are usually not effective in reducing demand charges. The installation of engine-driven generators is usually only economical where demand charges are very high and there is a specific short period

TABLE 12-3   Case Study: Wind-Diesel Hybrid System Purchased by a School

	No Wind Turbine	With 250-kW Wind Turbine	Wind Turbine + 75-kW Diesel Generator
Energy charge	$29,565	$15,768	$13,963
Demand charge	$45,000	$31,770	$18,000
Wind turbine loan payments	-	$38,611	$38,611
Wind turbine operating costs	-	$8813	$8813
Buyback revenue	-	($3066)	($3066)
REPI revenue	-	($2606)	($2606)
Diesel loan payments	-	-	$4726
Diesel operating costs	-	-	$4750
TOTALS			
During loan repayment period	$74,565	$89,289	$83,191
After loan is repaid	-	$50,678	$39,854

REPI = renewable energy production incentive.
(Source: Iowa Wind Energy Institute. 2001. *Wind hybrid electricity applications*. Iowa Department of Natural Resources. Des Moines, IA.)

of time when the customer's demand is measured. Running an engine-driven genera-tor during these short periods of time can be justified to reduce demand charges, even if the per-kWh operation costs of the engine-driven generator are high. Of course, if the engine-driven generator is needed for backup purposes anyway, then there may be instances where it would be economical to run it to reduce demand charges, even if the demand charge rates were not high.

In general, adding an engine-driven generator to a wind turbine installation will likely only be economical when the utility's demand charges are high or when there are additional benefits for having backup power for the customer, such as not having grid interconnection available at a remote site.

Transmission Constraints for Wind Hydro and Wind Diesel Hybrids

Large generation additions for or by utilities are increasingly being constrained by limitations on the bulk power transmission grid. The existing transmission grid has limited capacity to accept large amounts of additional generation in many areas, especially if they are some distance from the major load centers. Smaller generation additions, such as those discussed in this report, are not affected by these transmis-sion constraints because these smaller generation additions are more closely matched to the local load levels. However, these additions do not necessarily relieve any existing transmission constraints that may apply, for instance, to large wind farms.

BIODIESEL-FUELED WIND-DIESEL HYBRIDS

Hybrid wind-diesel engine energy systems are among the most popular hybrid renewable energy systems due to their relatively low capital costs and ability to provide constant power in the face of intermittent wind resources. Engine generator sets can be quickly started when wind speeds drop below levels required for wind-power generation.

The engine generator set provides a degree of flexibility in the choice of fuel burned. Most commonly diesel fuel is used, but this fossil fuel is a poor choice from a renewable energy perspective. However, there are three renewable fuels that can be used for standby power generation from a diesel engine: biodiesel oil, producer gas, and biogas. Biodiesel is manufactured from oils extracted from soybeans. Producer gas is the product of gasifying any of a number of biomass materials, including wood chips, switchgrass or other grass crops, and agricultural residues such as corn stover or corn cobs. Biogas is obtained from anaerobic digestion of high-moisture biomass such as manure.

Since biodiesel is chemically similar to diesel fuel burned in vehicles today, it can be used in the conventional diesel engines without degradation in engine performance or engine components. However, biogas and producer gas are both relatively low-grade fuels that contain contaminants that may be harmful to an engine. Therefore, spark ignition engines are often recommended for use with these fuels. A common diesel engine can be adjusted to burn a mixture of diesel and biogas or producer gas. Spark ignition engines are able to run on pure biogas or producer gas fuels. The rule-of-thumb maintenance cost for biogas and producer gas engines is $0.0125/kWh.

A variety of grains and seeds contain relatively high concentrations of vegetable oils that are of high caloric value and can be used as liquid fuels. However, research has shown that the high viscosity of these raw vegetable oils (20 times that of diesel) would lead to increased residues on the injectors and rings. On the other hand, chemical modification of vegetable oils to methyl or ethyl esters yields excellent diesel-engine fuel without the viscosity problems associated with raw vegetable oil. *Biodiesel* is the generic name given to these vegetable oil esters. Suitable feedstocks include soybean, sunflower, cottonseed, corn, groundnut (peanut), safflower, rapeseed, waste cooking oils, and animal fats. Waste oils or tallow (white or yellow grease) can also be converted to biodiesel.

In the production of biodiesel, the oil is squeezed from the seed in an oil press. By-product meal can be sold as a feed additive. The vegetable oil is "transesterified," a process by which triglycerides in the vegetable oil are reacted with methanol or ethanol to produce esters and glycerol.

The reaction is catalyzed by lye (NaOH or KOH) dissolved in 20% excess of methanol. The reaction proceeds rapidly to completion at room temperature in about an hour. Small amounts of soap are also produced by the reaction of lye with fatty acids. Upon completion, the glycerol and soap are removed in a phase separator. The heat of combustion of biodiesel is about 95% by weight of conventional diesel but, because it burns more efficiently, it generates essentially the same power as diesel fuel. Biodiesel can be used in unmodified diesel engines with no excess wear or operational problems. In over 100,000 km of tests using soybean oil methyl esters in light and heavy trucks, few significant problems arose other than the need for more frequent oil changes than conventional fuels because of the buildup of ester fuel in engine crankcases. Another advantage of biodiesel is that there is no net gain of CO_2 because the emission of CO_2 is equalized by the plants' consumption of CO_2. In addition, when biodiesel is burned there are less visible (particulate matter) emissions than the conventional diesel fuel. This helps keep the visible pollutants and greenhouse gases to a minimum. However, a disadvantage of biodiesel is the current high selling price, ranging between $2 to $5/gallon. It is not readily available in the United States. Below are nine sources of biodiesel in the United States:

- Ag Environmental Products, Lenexa, KS; 800-599-9209
- High Performance Fuels and Additives, Houston, TX; 713-467-7084
- Chemol Company, Greensboro, NC; 336-333-3071
- NOPEC Corporation, Lakeland, FL; 888-296-6732
- Columbus Foods, Chicago, IL; 773-265-6500

- Pacific Biodiesel, Kahului, HI; 808-877-3144
- CytoCulture, Point Richmond, CA; 510-233-0102
- World Energy Alternatives, Cambridge, MA; 617-868-1988
- Griffin Industries, Butler, KY; 703-256-4497

The use of biodiesel to run gensets as part of hybrid wind-diesel engine systems is not widely practiced today since conventional diesel fuel is very inexpensive in comparison.

PRODUCER GAS HYBRIDS

Producer gas is made by the conversion of solid, carbonaceous fuels into flammable gas mixtures. Producer gas consists of carbon monoxide (CO), hydrogen (H_2), methane (CH_4), nitrogen (N_2), CO_2, and smaller quantities of higher hydrocarbons. This gas can be burned directly in a furnace to generate process heat or it can fuel internal combustion engines, gas turbines, or fuel cells.

In a fluidized bed gasifier, a gas stream passes vertically upward through a bed of inert particulate material to form a turbulent mixture of gas and solid. Fuel is added at such a rate that it is only a few percent by weight of the bed material. The violent stirring action makes the bed uniform in temperature and composition with the result that "gasification" (liquid fuel becoming vaporized) occurs simultaneously at all locations in the bed. Fluidized beds are gaining favor as the reactor of choice for biomass gasification. The high thermal mass of the bed imparts a high degree of flexibility in the kinds of fuels that can be gasified, including those of high moisture content. Disadvantages include relatively high exit gas temperatures, which complicate efficient energy recovery, and relatively high amounts of impure particles in the gas due to the abrasive forces acting within the fluidized bed.

Although gasifier-engine systems have been demonstrated, they can be difficult to operate, are only moderately reliable, and have lifetimes of only 10 years. They also pose a potential threat of CO poisoning. Many gasification systems require constant adjustments during operation. Although automatic controls can be employed in their operation, hybrid wind-gasifier-engine systems are probably not practical at this time for homeowners or farmers because of their complexity.

Producer gas hybrids are better suited for industrial purposes where economies of scale and full-time personnel are available. They would, however, be more suited for industrial or utility purposes where full-time personnel are available to attend them. Additionally, if a company generated biomass waste products the fuel costs would drastically decrease, thus making gasification more economically viable. Larger-scale systems also capture important economies of scale, which reduces the cost of electricity as the size of the system increases. Figure 12-5 shows a typical hybrid wind-gasifier system.

According to David Walden of Bronzeoak Corporation, the cost of a downdraft gasifier sized at 2 MW coupled with a "diesel" genset adapted to operate on 70% producer gas and 30% diesel would cost around $1300 per kW (gross). This is an unusual approach. More commonly, the spark ignition engine is used to burn 100% producer gas. For this approach, the total costs of the system could be around $2000 per kW (gross).

A resource for manufacturers of large- and small-scale gasifiers can be found in "A Survey of Biomass Gasification 2000: Gasifier Projects and Manufacturers Around the World" published by the National Renewable Energy Laboratory. This source is an international directory of gasifier manufacturers. It includes schematics, cost information, company history, current projects, and additional references with each manufacturer.

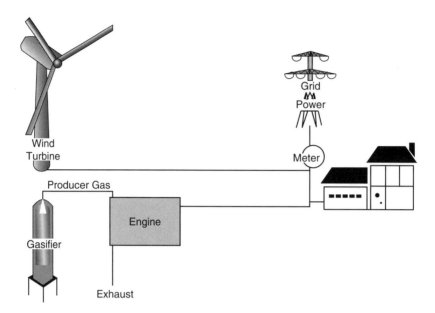

FIGURE 12-5 *Hybrid wind-gasifier system. (Source: Iowa Wind Energy Institute. 2001.* Wind hybrid electricity applications. *Iowa Department of Natural Resources. Des Moines, IA.)*

BIOGAS HYBRIDS

Biogas is distinguished from producer gas in that it is made through anaerobic (oxygen-free) digestion of organic waste to gaseous fuel by bacteria. The process occurs in stages to successively break down the organic matter into simpler organic compounds. The product is a mixture of CH_4, CO_2, and some trace gases. Most digestion systems produce biogas that is between 55% and 75% CH_4 by volume. Biogas, once treated to remove sulfur compounds, can be used in many applications, including stationary power generation. Biogas is a suitable fuel for engine generator sets, small gas turbines, and some kinds of fuel cells. The biological processes within an anaerobic digester that lead to biogas are relatively complicated. The process consists of three basic steps: hydrolysis and fermentation, transitional acetogenic dehydrogenation, and methanogenesis. Hydrolysis and fermentation involve hydrolytic and fermentative bacteria that break down proteins, carbohydrates, and fats into simpler acids, alcohols, and neutral compounds. Hydrogen and CO_2 are also produced. This step is followed by transitional digestion through the action of acid-forming bacteria. Products of fermentation that are too complex for CH_4-forming bacteria to consume are further degraded in this step to acetate, hydrogen, and CO_2. Traces of oxygen in the feedstock are consumed in this step, which benefits oxygen-sensitive, CH_4-forming bacteria. The final step, methanogenesis, converts acetate to CH_4 by the action of CH_4-forming bacteria.

Today anaerobic digesters are used to produce CH_4, which can be used to produce electricity by operating a genset adapted for CH_4. The alternating current (AC) output from the genset can be simply tied into an inverter and sent through a DC bus. An anaerobic digestion system can be combined with a wind turbine, as illustrated in Figure 12-6.

Basic anaerobic digestion systems are relatively simple. For some feedstocks, removal of scum may be desirable in advance of the anaerobic digester. Whether batch or steady-flow processing, two effluent streams result: biogas and sludge. The biogas

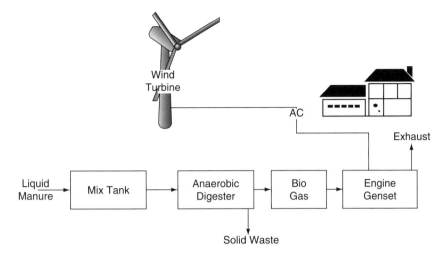

FIGURE 12-6 *Hybrid wind-anaerobic digester. AC = alternating current. (Source: Iowa Wind Energy Institute. 2001.* Wind hybrid electricity applications. *Iowa Department of Natural Resources. Des Moines, IA.)*

may have to be scrubbed in advance of combustion to remove H_2S that would otherwise appear as the pollutant sulfur dioxide in the exhaust gas. H_2S increases concerns about wear to engine parts and bearings, which causes maintenance issues.

Examples of anaerobic reactor designs include the simple batch reactor, plug-flow reactors, continuously stirred tank reactors, upflow reactors, and two-tank reactor systems. The batch reactor is a single vessel design in which waste is added and removed in batches and all steps of the digestion process take place in the same environment. More advanced reactor designs aim to improve waste contact with active bacteria and/or to separate and control the environments for acid-forming and CH_4-forming bacteria. Relatively new two-tank reactor technology (sometimes called a *sequence-based reactor*) physically separates the acid formation and CH_4 formation phases of digestion so that each takes place under optimal conditions. Benefits of this technology are numerous: hydrolysis and acidification occur more quickly than in conventional systems; the common problems of foaming in single-tank systems are reduced by the destruction of biochemical foaming agents before they reach the CH_4-forming reactor; and the biogas produced is typically rich in CH_4.

The biological conversion process of anaerobic digestion can be used for a wide range of feedstocks of medium-to-high moisture content. Sewage sludge is most commonly used as feedstock for anaerobic digesters, but municipal solid waste, food-processing wastes, agricultural wastes, Napier grass, kelp, bagasse, and water hyacinth can also be digested. Digesters are designed to maintain optimal conditions for a specific type of waste. Waste pretreatment, heating, mixing, nutrient addition, specialized bacteria addition, and pH adjustment can be manipulated to control digester performance.

Typical thermodynamic efficiency in converting the chemical energy of dry matter to CH_4 is about 60% with CH_4 yields ranging between 8000 and 9000 cubic feet per ton of volatile solids added to the digester. However, thermodynamic efficiencies approaching 90% with CH_4 yields of 11,000 to 13,000 cubic feet per ton of volatile solids have been achieved in research programs.

When compared with the conventional aerobic digestion there are numerous advantages which include less solid waste, less odor, reductions in fly population, less space taken up by lagoons, and valuable by-products in the form of fertilizer. In addition, anaerobic digesters help reduce the amount of greenhouse gases (CH_4) that are released into the atmosphere with conventional aerobic digestion techniques. Some disadvantages are that anaerobic digesters require regular maintenance and, in the past, many digesters have failed due to bad designs. Design of anaerobic digesters is solely dependent upon the farmer's needs and costs accordingly. Typical payback can range up to 10 years. These digesters would be ideally suited economically for large-scale dairy, swine, and poultry operations where high wastes are available. A skilled operator would also be beneficial for daily maintenance and problems.

There are CH_4-recovery projects in Nevada and Creston, IA. One operation converts manure from a 4800-animal farrow-to-wean operation into almost 912 thousand cubic feet of biogas per month. The biogas is burned in a genset that generates 67,240 kWh per month. There are also three 250-kW wind turbines in one of the projects in Nevada, but they are not directly interconnected as a hybrid system such as proposed in Figure 12-6.

SOLAR HYBRIDS

Many wind hybrid systems incorporate solar energy. Solar energy can be harnessed through a number of means. Photovoltaic panels and materials convert sunlight directly to electricity through a photovoltaic chemical process. Solar thermal panels collect the sun's heat to heat water and thereby reduce the energy requirement of water heating in a residence. Double-pane windows also trap the sun's heat in the winter, reducing heating-energy requirements in the home. Any method of conserving energy or reducing the energy load will improve the efficiency of a renewable system. For the purposes of generating electricity using a wind hybrid system, this section of the report will focus on the merits of incorporating photovoltaics.

The photovoltaic technology that uses crystalline silicon is used in over 90% of current installed photovoltaic capacity. More efficient photovoltaic materials are available but at a much higher price. Photovoltaic technology is characterized by an increase in efficiency as the temperature falls (hence arrays may deliver more than their rated output in winter). Some photovoltaic systems can make use of all available solar energy (net of system losses), such as grid-connected systems or water-pumping systems. Stand-alone residential systems with batteries are the most common type of solar installation. These systems often integrate a wind turbine and backup generators to create a complete hybrid energy system. Having a backup power supply or batteries with a photovoltaic system that is not grid-interconnected is essential to be able to provide energy to meet load requirements when the sun is not shining. However, the incorporation of batteries reduces the overall efficiency of the system compared to grid-interconnected systems or a system sized so that all the energy is directly used.

When estimating the output of a photovoltaic system, losses need to be included. Inverter losses are taken into account intrinsically by considering an AC rating for the photovoltaic system. Some photovoltaic systems included trackers that rotate the photovoltaic panels to continually face the sun as it moves. Many tracking systems rely on passive trackers that do not consume energy. Active trackers would use about 1% of the energy produced. Battery losses would be relevant only for systems that are not grid connected. Small monitored systems connected to batteries have recorded 30% of the photovoltaic energy lost through: (1) round-trip battery storage, and (2) overproduction at times with small loads and full batteries. This 30% would be in addition to inverter losses. All together, the average losses on a system connected to batteries with an inverter are on the order of 40% to 45%.

Photovoltaics are usually incorporated into solar panels that can be mounted on a roof or stand. Some stands are designed to track the sun, increasing the efficiency of the solar panel. Home-scale photovoltaic panels often use battery storage and a backup diesel or propane generator to provide power when sunlight is not available.

MICRO-HYDROELECTRIC HYBRIDS

Where flowing water exists, it may be possible to generate electricity with a micro-hydroelectric turbine. Hydroelectric generation, where feasible, has a very high return over its lifetime. Many of these systems can provide a relatively constant power supply. Most commonly used in "home" applications are impulse turbines. This device uses either a Pelton wheel or a Turgo wheel to turn flowing water into AC electricity. The Turgo wheel is a rugged bronze turbine wheel that is applicable to a wide range of elevation heads as small as 2 meters. The Pelton wheel is most useful for higher-head, lower-flow situations. Both types of turbines must be supported by a "penstock" that channels flowing water into a pipe, which then feeds into the turbine and turns the wheel. All small hydroelectric turbines have fairly high losses that limit energy conversion efficiencies to 25% to 50%. Losses arise within the turbine as well as from friction between the pipe and moving water.

These losses must be accounted for when evaluating power requirements. A rectifier, which turns AC to DC, is often a standard on most turbines since battery storage is assumed. If the turbine is used for a home, an inverter is required to change the current back to AC. *Micro-hydroelectric* refers to turbines that are able to operate under low-elevation-head or low-volumetric-flow-rate conditions. Elevation head and volumetric flow rate determine the water pressure and amount of water flowing and are therefore important factors in determining the feasibility of micro-hydroelectric power for a particular location. It is these conditions that determine the size of the micro-hydroelectric turbine to be used since each turbine is dependent upon the flow and head. The U.S. Department of Energy maintains a Web site that is helpful in determining elevation head, volumetric flow rates, legal requirements, costs, and other considerations in constructing a successful micro-hydroelectric system for home application at http://www. eren.doe.gov/consumerinfo/refbriefs/ab2.html.

In the Midwest, where winters are very cold and summers are often dry, streams and some rivers usually freeze entirely in the winter or dry up in the summer. In these environments it may be necessary to couple a micro-hydroelectric turbine to a wind turbine or photovoltaic array to ensure a constant power supply. Figure 12-7 schematically illustrates a hybrid wind/hydro/battery system.

On the other hand, if the water resource is plentiful year round, a hydroelectric turbine with a battery bank could provide an adequate supply of electricity without the addition of a wind turbine or photovoltaic array. Accordingly, a careful assessment of hydraulic resources should be performed before deciding to incorporate it into a hybrid energy system. The typical cost of a two-nozzle Harris Pelton Turbine that produces 1400 watts at 24 volts will cost about $1000 without installation. Installation is the major capital cost of a hydro system, costing between $5000 and $10,000.

Another way to generate electricity from relatively fast-flowing rivers or streams is the use of submersible generators. The Aquair UW 100-watt Submersible generator is one of several commercially available units. The generator consists of a simple propeller attached to a shaft that turns an alternator sealed in a metal housing filled with oil. The output is AC power that can be converted to DC power by a suitable rectifier for storage in electric batteries (12 or 24 volt). Figure 12-8 illustrates a submersible generator coupled to a wind turbine and battery-charging system.

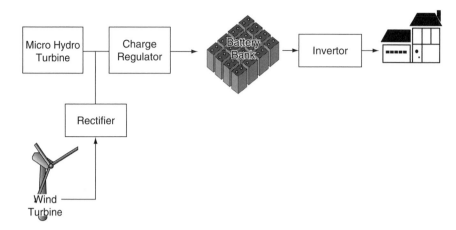

FIGURE 12-7   *Hybrid wind/hydro/battery system. (Source: Iowa Wind Energy Institute. 2001. Wind hybrid electricity applications. Iowa Department of Natural Resources. Des Moines, IA.)*

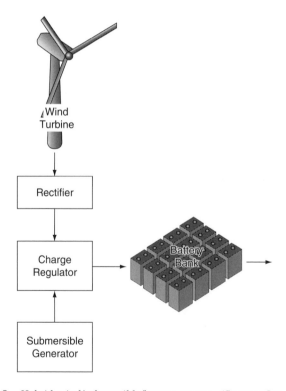

FIGURE 12-8   *Hybrid wind/submersible/battery system. (Source: Iowa Wind Energy Institute. 2001. Wind hybrid electricity applications. Iowa Department of Natural Resources. Des Moines, IA.)*

This system is more benign to the environment than micro-hydroelectric turbines, because the submersible generator does not require any special pipelines or modifications to the stream. The submersible generator is simply lowered into a flowing stream. They can be mounted to bridges via a vertical support that allows adjustments to obtain suitable depths in a seasonal stream. Stream channeling can be employed to increase flow velocities with subsequent increases in electrical power, but is not necessary. The cost of a submersible turbine is low but the output power is also low compared to micro-hydroelectric turbines. At water speeds between 2.5 and 4.0 meters per second the power output ranges between 60 and 100 watts. The Aquair submersible generator requires only 0.33 meters (13 inches) of water depth to operate. The cost of the Aquair submersible generator is about $1300, which is an expensive $13,000/kW of capacity. Oasis Montana is one of many dealers that sell the Aquair 100-watt submersible generator. As with the micro-hydroturbine systems, seasonal flow variations may cause problems.

IN GENERAL

Despite aggressive wind turbine proactivity in Europe, the world is far from tapping its wind potential, which could be 20% to 30% of global consumption.

Two percent of the world's incident solar energy can fulfill the entire world's power needs. The world's population is not quite ready to break with convention, however, and must take the slow road to "clean air." The microturbine is one of the steps on that slow road.

It is unlikely that any distributed energy source will succeed without government assistance. In the United States, Canada, and Europe, this help may be available at the federal, state, or provincial level and regional levels.

The U.S. Department of Energy (DOE) is one of several government organizations tasked with helping distributed energies attain their potential. Like all government agencies, everything it develops and whatever is developed using its funding is made publicly available. The DOE attempts to support all of these different energy sources to the extent it deems appropriate, given its funding level. The importance of DOE support cannot be overrated. So, its summary of its Distributed Energy Programs support is quoted in Annex 1 of this chapter. More information can be collected by visiting the DOE Web site.

ANNEX 1

The U.S. Department of Energy's Distributed Energy Program* was established in fiscal year 2001. The program develops a portfolio of advanced, on-site, small-scale, modular energy conversion and delivery systems for industrial, commercial, residential, and utility applications.

Program activities are organized under two main areas: Distributed Generation Technology Development and Integrated Energy Systems.

Distributed Generation Technology Development

This effort seeks to develop a portfolio of electricity generation and heat utilization technologies with a focus on efficiency, emissions, RAMD (i.e., reliability, availability, maintainability, and durability), and meeting cost targets. By improving the efficiency

---

* Source: U.S. DOE public information Web site. Available at http://www.energy.gov.

of thermally activated systems and advancing the efficiency and emissions characteristics of these power-generation technologies, the program provides the building blocks necessary to develop advanced, integrated systems.

Integrated Energy Systems

The focus of this effort is to develop highly efficient integrated energy systems that can be replicated across end-use sectors and that will help demonstrate a research-and-development objective or address a technical barrier.

The Distributed Energy Program supports the development of the next generation of clean, efficient, reliable, and affordable distributed energy technologies and the integration of these technologies at end-user sites. The program works with small, modular power–generating technologies that can be combined with energy management and storage systems to improve the operation of the electricity delivery system.

Distributed energy technologies are defined as small-scale, modular technologies for on-site, grid-connected, or stand-alone energy conversion and delivery. They include the following:

- Gas-fired reciprocating engines
- Industrial gas turbines
- Microturbines
- Technology-based research
- Thermally activated technologies

Integrated energy systems are systems that combine distributed power generation with equipment that uses thermal energy to improve overall energy efficiency and fuel use. They include the following:

- CHP applications
- CHP technologies

The U.S. DOE's Distributed Energy Program is working with utilities, energy service companies, industrial manufacturers, and equipment suppliers to identify technologies that will improve the energy, environmental, and financial performance of power systems for manufacturing, processing, and other commercial applications. The program will contribute to the development of ultrahigh-efficiency and low-emission engine systems and provide new choices and innovative power solutions to the industrial sector.

Chapter 13

Microturbines in Integrated Systems, Fuel Cells, and Hydrogen Fuel

As outlined in Chapter 12, microturbines are also used in conjunction with other distributed energy sources to form an integrated system. These systems offer better stability, greater operating range, and reliability, as well as increased efficiency, than either energy method would on its own.

In this chapter, we will also examine fuel cells, as they offer efficiencies as high as 80% independently, and in conjunction with microturbines. Their drawback, as we will see, is their high cost in terms of dollars per kilowatt (kW). Also, their optimum fuel is hydrogen, and the manufacture of hydrogen is another item that needs to drop in cost to be viable on the commercial market.

Microturbines on their own without any waste-heat recovery (no recuperator) are typically about 25% efficient. If they have a recuperator, efficiency climbs to 40%.

Microturbines in conjunction with a fuel cell have produced systems with 80% efficiency. This is a promising application, despite the high capital cost of fuel cells. In areas where there is no grid, even an expensive fuel cell on its own could provide a return on investment in about 2 years if fuel cost were high enough (see Chapter 12).

Recuperators can be expensive to manufacture, as one might expect with a component that must include a large amount of heat exchange capacity in a relatively small space. Cogeneration applications can include waste-heat recovery methods that do not need a recuperator. Direct use of the microturbine's exhaust gas heat can produce some excellent efficiency figures, even without another separate distributed energy source.

Such cases are illustrated by, for instance, work by a group of participants including the Oak Ridge National Laboratory (paid for by parties that include the federal Office of Distributed Energy Resources) that used waste microturbine heat in a desiccator, used for instance, to dry paint (Figure 13-1). The resultant system efficiency was on the order of 70%. The heat can be used to heat water as well. The Oak Ridge National Laboratory is working with Georgia Tech Research Institute and other partners to monitor the effects of continuous ventilation from—and exposure to the water vapor levels in—a microturbine's exhaust gases. This could add a new facet of applications in HVAC systems.

Installations in nine rural communities (Table 13-1) are the test base. One of them had a Honeywell 75-kW microturbine. Capstone Turbines, in California, supplied some of the other turbines.

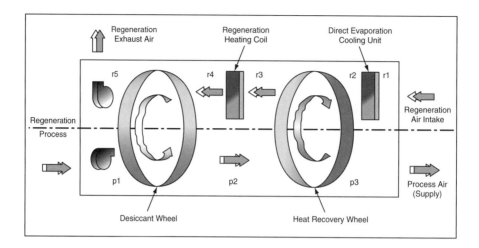

FIGURE 13-1   *Schematic of the operation of a desiccant dehumidification system. (Source: Office of Power Technologies. Desiccant technology keeps indoor air fresh, healthy, comfortable.)*

TABLE 13-1   Microturbine Field Tests at the Nine Rural Cooperatives

Cooperative	Location	Microturbine	Application	Thermal recovery
AVEC	Anchorage, AK	30-kW Capstone	Power at warehouse and office	Space heating
Arkansas Electric	Augusta, AR	75-kW Honeywell*	Power at electricity-generating facility	Freeze protection for plant boiler
Associated Electric	Kearney, MO	45-kW Elliott	Power for city water well pumps	TBD
Cass County Electric	Fargo, ND	30-kW Capstone	Power for domestic water heating at hotel and convention center	Preheat hot water for gas-fired boilers
Chugach	Anchorage, AK	30-kW Capstone	Power at warehouse and office	TBD
East Kentucky Electric	Winchester, KY	60-kW Capstone	Power at facility electricity-generating	TBD
Kotezebue Electric	Kotezebue, AK	45-kW Elliott or 60-kW Capstone	Power plant and later a commercial customer	TBD
Tennessee Valley Authority	Muscle Shoals, AL	30-kW Capstone	R&D office building	None
Tri-State G&T	Brighton, CO	70-kW Ingersoll-Rand	Industrial greenhouse	Preheat plant irrigation water

TBD = To be decided.
*Honeywell is closing its microturbine manufacturing facilities and will buy back purchased units, so this unit will be replaced with a microturbine from another manufacturer.

Microturbines used in conjunction with fuel cells are raising a great deal of interest. The work on a test supported by the European Economic Community government follows.*

1-MW SOLID OXIDE FUEL CELL—HYBRID FUEL CELL/MICROTURBINE SYSTEM

Challenges

The program addressed the design, manufacture, and operation of a natural gas–fueled pressurized hybrid solid oxide fuel cell (SOFC)/microturbine generator power system (Figure 13-2). The specific issues to be addressed, that will contribute to advancing the technology for SOFC power systems in Europe, include:

- Designing the SOFC module for MW-class systems and pressurized SOFC operation
- Availability of a recuperated microturbine generator that can satisfy the pressure, air inlet temperature, and airflow characteristics required for direct integration with the pressurized SOFC module
- European supplier development to maximize the European content in the pressurized hybrid power systems

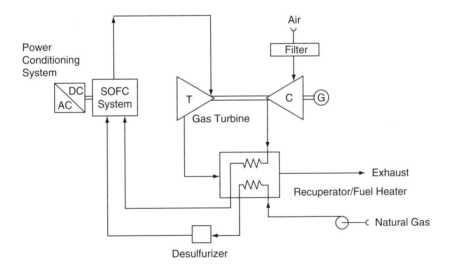

FIGURE 13-2 *Process schematic for a simple pressurized solid oxide fuel cell (SOFC) gas turbine combined-cycle concept. AC = alternating current; DC = direct current; T = turbine; C = compressor. (Source: Hanssen, J. E. 1MW SOFC hybrid fuel cell/micro-turbine system.* Demonstration of a MEeI class power system using high temperature fuel cells [SOFC] combined with micro-turbine generators [1MWSOFC]. *European Economic Community.)*

---

* Source: Hanssen, J. E. 1 MW SOFC hybrid fuel cell/micro-turbine system. *Demonstration of a MEeI class power system using high temperature fuel cells (SOFC) combined with micro-turbine generators (1MW SOFC).* European Economic Community.

- Module cost reduction and system design simplification to reduce cost of the commercial prototype pressurized hybrid system
- Cost-effective, skid-mounted component packaging
- Satisfying European code and safety requirements

The main technical targets of the proposed program are:

- Operation of the first MW-class SOFC power system of European design and containing significant European content
- System efficiency >55% (approaching 60%) for the pressurized hybrid system

Project Structure

The project consortium consists of Electricité de France (EDF), Energie Baden-Württemberg (EnBW, Germany), Gaz de France (GDF), Siemens AG (SAG, Germany), Siemens Westinghouse Power Corporation (SWPC, USA), and Tiroler Wasserkraft AG (TIWAG, Austria). SWPC as advisor is reporting to the U.S. Department of Energy (DOE), EnBW as project coordinator is reporting to the EC.

The main work packages are:

- Product specifications
- SOFC module
- System engineering/design/assembly
- Microturbine generator/thermal management system
- Site preparation
- Installation, shakedown, and acceptance
- System test

Work on this project continues.

MICROTURBINE-WIND HYBRIDS

Microturbines, in conjunction with wind turbines, have been studied by the government of the state of Iowa. The primary focus for their study, excerpts of which are included both in this and the previous chapter, was their natural resources of wind and hydro (i.e., river dams). The main topic of discussion was wind hybrids (see Chapter 12). However, the experience cited for microturbine and fuel cell hybrids (with wind) was significant, hence its inclusion here.

Microturbine-Wind Hybrid Case*

When burning natural gas in a microturbine, NO_x emissions are only 0.1 to 0.5 lb per megawatt-hour (MWh) compared to 4.0 to 10.0 lb/MWh for coal-fired power plants and 0.1 to 2.0 for a large combustion turbine. Generally, microturbines are manufactured with commercial applications in mind, but a 25-kW system would be able to supply continuous electricity to 10 to 12 homes, assuming a 2-kW consumption per house. This would be ideal for a small community that was isolated from a utility grid.

Typical installation costs of microturbine systems range from $450 to $700/kW. This is currently an expensive alternative to diesel gensets. However, the benefits

---

* Source: Adapted from Iowa Wind Energy Institute. 2001. *Wind hybrid electricity applications.* Iowa Department of Natural Resources. Des Moines, IA.

and anticipated rapid declines in price as mass production begins should be weighed in the consideration.

Even though microturbines are designed to continuously supply power to homes and farms, they may merit consideration for hybrid wind energy systems as well. The microturbine can be run on biodiesel or ethanol as a completely renewable energy option for backup power. The cost of energy from the wind turbine would be much less than the microturbine and save on fuel and operating costs.

Significantly, the electric power industry frequently uses large gas turbines as part of emergency or peak power capacity. Typical start-up times range from 20 to 120 seconds. A simple battery bank may eliminate transients in power as a system switches from wind to microturbine; however, microturbines were not designed for repeated start-and-stop cycles.

Companies are trying to modify conventional microturbines to burn lower-quality fuels such as biogas and producer gas from anaerobic digestion and gasification, respectively. These fuels cause problems in current microturbines, because microturbines require a very clean gas. Impurities in the gas stream that are commonly found in biogas and producer gas can damage turbine blades and reduce the life of the system.

Hydrogen obtained from electrolysis of water using wind turbine electricity could be envisioned as part of an energy storage and recovery system, although this application is probably several years away. The relatively high capital cost of microturbines reduces its attractiveness, though, in applications where it is on standby for long periods.

The Iowa study's cost data could have been split between the previous chapter and this one, but for easier reference, they follow in this chapter.

Technology Cost Comparison[*]

The purpose of this section is to show the relative cost per kilowatt-hour of the different technologies described in this report. This report focuses on the home and commercial-scale units available on the market today. A full hybrid system may also contain a controller that switches between the various generators and storage according to the load requirement. An example of a hybrid system with batteries, generator, wind generator, and controller was given in Chapter 12 in the section titled "Solar Hybrids."

For combustion generators, the assumptions on fuel costs are listed in Table 13-2.

When considering the wind turbine for the system, the price per kilowatt drops considerably as the size of the turbine increases. In today's market, small turbines are available in sizes such as 500 watts (W), 1000 W, 3000 W, 10 kW, 20 kW, and 50 kW. The price per kilowatt of these turbines with the necessary tower and transformation is at least $1500/kW. Large turbines range in size from 600 kW to 2.5 MW. When installed in a wind farm, they achieve economies of scale, bringing the price down to $850 to $1250/kW. Large turbines are usually more efficient as well.

Below are five steps to evaluating the cost benefit of a wind hybrid system and its components. These steps may be followed sequentially, or the reader may wish to skip to the information of interest.

Step One: Determine Energy Needs and Current Costs

Here, the reader should determine the amount of energy that is typically used or desired from the wind hybrid system. By looking at utility bills, the cost being currently paid

---

[*] Source: Iowa Wind Energy Institute. 2001. *Wind hybrid electricity applications*. Iowa Department of Natural Resources. Des Moines, IA.

TABLE 13-2   Cost Assumptions for Combustion Generators

Natural Gas		*Kerosene*	#1 diesel
Density	0.79 kg/m³	Density	840 kg/m³
Heating value	45,000 kJ/kg	Heating value	43,200 kJ/m³
Cost per energy	**0.0038 $/MJ**	**Cost per energy**	**0.0117 $/MJ**
Fuel cost	$0.13/m³	Fuel cost	$425.31/m³
Propane (Liquid)		*Diesel*	#2 diesel
Density	500 kg/m³	Density	880 kg/m³
Heating value	46,357 kJ/kg	Heating value	42,800 kJ/kg
Cost per energy	**0.0193 $/MJ**	**Cost per energy**	**0.0119$/MJ**
Fuel cost	$446.45/m³	Fuel cost	$446.45/m³
Gasoline		*Biodiesel*	
Density	720 kg/m³	Density	875 kg/m³
Heating value	44,000 kJ/kg	Heating value	37,200 J/kg
Cost per energy	**0.0108$/MJ**	**Cost per energy**	**0.0203$/MJ**
Fuel cost	$343.42/m³	Fuel cost	$660.43/m³
Switchgrass		*Corn Stover*	
Yield	3628.74 kg/acre	Yield	1817.36 kg/acre
Bulk density	130 kg/m³	Bulk density	130 kg/m³
Heating value	18,004.61 kJ/kg	Heating value	17,444.00 kJ/kg
Cost per energy	**0.0031 $/MJ**	**Cost per energy**	**0.0047 $/MJ**
Fuel cost	$201.94/acre	Fuel cost	$150.00/acre

(Source: Iowa Wind Energy Institute. 2001. *Wind hybrid electricity applications.* Iowa Department of Natural Resources. Des Moines, IA.)

for electricity should be recorded. This information will be carried forward to the subsequent steps for comparative analysis.

a. Average annual electric usage in kilowatt-hours
b. Average annual cost of electricity
c. Average cost per kilowatt-hour (divide b by a)

Step Two: Determine Wind Resources Output

The Iowa Energy Center Web site (http://www.energy.iastate.edu/wind/calc.cfm) provides the ability to look up the location of the proposed wind turbine site and then to select a turbine size and tower height appropriate for the site and energy requirement. The result is a calculation of the proposed turbine's monthly and annual energy output along with the average wind speeds and turbine capacity factor (a measure of the turbine's efficiency).

Step Three: Cost Comparison of Wind/Hybrid Components

The third step is to integrate the information in the first two steps and add hybrid components (other energy sources or energy storage) to the system. Using Table 13-3 or logging onto the hybrid cost comparison calculator Web site of the Iowa Department of

TABLE 13-3 Cost Comparison Calculator for Large and Small Wind Turbines

Large Wind Turbine		Small Wind Turbine	
Fuel Cost		*Fuel Cost*	
Average capacity factor	30.00%	Average capacity factor	25.00%
Rated power output	900 kW	Power output	20 kW
Fuel conversion efficiency	na	Efficiency	na
Fuel consumption	0 MJ/kW	Fuel consumption	0 MJ/kW
Fuel cost per energy	$0.00000/MJ	Fuel cost per energy	$0.00000/MJ
Annual fuel cost	*$0/year*	*Annual fuel cost*	*$0/year*
Cost of Capital		*Cost of Capital*	
Rated capacity	900 kW	# of kilowatts (nameplate)	20 kW
Est. price per kW	$1,000/kW	Est. price per kW	$1500/kW
Period of loan	15 years	Period of loan	8 years
Amount of loan	$900,000	Amount of loan	$30,000
Interest rate of loan (annual)	10.00%	Interest rate of loan (annual)	10.00%
Amt. paid over life of loan	$1,774,896	Amt. paid over life of loan	$44,987
Payments per year	*$118,326/year*	*Payments per year*	*$5623/year*
TOTAL	**$118,326**/*year*	**TOTAL**	**$5623/year**
Cost per kilowatt-hour	*$0.128 $/kWh*	*Cost per kilowatt-hour*	*$0.050 $/kWh*
kWh produced annually	*2,365,200 kWh*	*kWh produced annually*	*43,800 kWh*

(Source: Iowa Wind Energy Institute. 2001. *Wind hybrid electricity applications.* Iowa Department of Natural Resources. Des Moines, IA.)

Natural Resources: http://www.wuebdesign.com/energy/pubs/whea/analysis/index.htm. This Web site allows the user to compare kilowatt-hour price for different hybrid components. The capacity factor and wind turbine size determined from step 2 are entered here. An opportunity is provided to adjust the default fuel costs if using a generator or biomass with fuel locally available. The result is a cost per kilowatt-hour for each component and the system as a whole. These costs are for comparison's sake only. They only include equipment purchase and fuel costs. They do not include installation, operation, maintenance, warranty, or "life-cycle" costs.

If the hybrid system includes solar photovoltaics, refer to the following Web site for more information on the energy-output characteristics of photovoltaics at a chosen site in Iowa: http://www.energy.iastate.edu/solarcalculator/index.

Wind "capacity factor" refers to the average percentage of the turbine's rated capacity that the turbine achieves in Iowa winds. Table 13-4 provides a definition of the terms used in the analysis of the different generators in Table 13-5.

FUEL CELLS

Fuel cells are a viable power-generation source on their own, and potentially the cleanest distributed energy source and possibly the commonest method by about the middle of the century. So their "solo" application deserves some discussion, as does hydrogen fuel. Those two items follow.

TABLE 13-4　Terms Used in Generator Analysis: Definitions

Fuel Cost	Definitions
Percent of time in operation	*Capacity required to offset wind capacity factor*
Rated power output	*Maximum output available on typical-sized unit*
Fuel conversion efficiency	*Efficiency of converting fuel to energy*
Fuel consumption	*Fuel consumed to produce 1 kilowatt*
Fuel cost per energy	*Fuel cost of type of fuel required*
Annual fuel cost	*Total fuel cost for the period*
Cost of capital	
Average capacity factor	*Average percent of rated capacity delivered (net)*
Rated capacity	*Rated maximum power output of unit*
Est. price per kW	*Cost to produce 1 kilowatt*
Period of loan	*Length of loan financing*
Amount of loan	*Average purchase cost of unit*
Interest rate of loan (annual)	*Loan interest rate*
Amt. paid over life of loan	*Cumulative payment at loan completion*
Payments per year	*Payment for the period*
TOTAL	*Total for the period*
Cost per kilowatt-hour	*Cost per kilowatt hour*
Total kWh/year	*Total kilowatt hours delivered per year*

(Source: Iowa Wind Energy Institute. 2001. *Wind hybrid electricity applications.* Iowa Department of Natural Resources. Des Moines, IA.)

A fuel cell[*] is an electrochemical device that converts the chemical energy of hydrogen into electricity, heat, and water vapor. This conversion can be done with high electrical efficiency (35% to 55%) and with minimum environmental effects. This could make fuel cells the favorite energy-conversion d evices in the medium (after 2040) to long term, in both transport and stationary applications and for all power ranges. Fuel cells operate with a variety of fuels, including biofuels.

The fuel cell process is electrolysis in reverse. In an electrolyser, electric current is passed through water, which is split into oxygen and hydrogen. In a fuel cell hydrogen and oxygen are combined, producing an electric current and water.

Fuel cells operate like batteries, except that in batteries the reactants are stored within the battery itself and are limited by its size. In a fuel cell they are stored externally and energy can be produced as long as fuel is fed to the anode and an oxidant to the cathode.

In a fuel cell the hydrogen fuel reacts at the anode in the reaction:

$$2H_2 + O_2 \rightarrow 2H_2O$$

Electrical energy and heat are produced. This electrical energy is:

- Proportional to the contact area between the gases, the electrolyte, and the electrodes, and
- Inversely proportional to the distance between the electrodes.

---

[*] Source: N. Lymberopoulos, N. 2005. *Fuel cells and their application in bio-energy.* Project Technical Assistant Framework Contract (EESD Contract No: NNE5-PTA-2002-003/1).

TABLE 13-5   Analysis of Different Types of Generators

Diesel Engine		Microturbine	
Fuel Cost		*Fuel Cost*	
Percent of time in operation	75%	Percent of time in operation	75.00%
Rated power output	25 kW	Rated power output	25 kW
Fuel conversion efficiency	25%	Fuel conversion efficiency	25.00%
Fuel consumption	100 MJ/kW	Fuel consumption	100 MJ/kW
Fuel cost per energy	$0.01185/MJ	Fuel cost per energy	$0.00379/MJ
Annual fuel cost	*$28,036/year*	*Annual fuel cost*	*$8962/year*
Cost of Capital		*Cost of Capital*	
Rated capacity	25 kW	Rated capacity	25 kW
Est. price per kW	$500/kW	Est. price per kW	$700/kW
Period of loan	8 Years	Period of loan	8 years
Amount of loan	$12,500	Amount of loan	$17,500
Interest rate of loan (annual)	10.00%	Interest rate of loan (annual)	10.00%
Amt. paid over life of loan	$18,744	Amt. paid over life of loan	$26,242
Payments per year	*$2343/year*	*Payments per year*	*$3280/year*
TOTAL	**$30,379/year**	**TOTAL**	**$12,242/year**
Cost per kilowatt-hour	*$0.185 $/kWh*	*Cost per kilowatt-hour*	*$0.075 $/kWh*
kWh produced annually	*164,250 kWh*	*kWh produced annually*	*164,250 kWh*
Submersible Generator		Hydroturbine (micro)	
Fuel Cost		*Fuel Cost*	
Percent of time in operation	75.00%	Percent of time in operation	75.00%
Power output	0.5 kW	Rated power output	1.4 kW
Efficiency	na	Fuel conversion efficiency	na
Fuel consumption	0 MJ/kW	Fuel consumption	0 MJ/kW
Fuel cost per energy	$0.00/MJ	Fuel cost per energy	$0.00/MJ
Annual fuel cost	*$0.00/year*	*Annual fuel cost*	*$0.00/year*
Cost of Capital		*Cost of Capital*	
Rated capacity	0.5 kW	Rated capacity	1.4 kW
Est. price per kW	$13,000/kW	Est. price per kW	$6071/kW
Period of loan	8 years	Period of loan	8 years
Amount of loan	$6500	Amount of loan	$8499
Interest rate of loan (annual)	10.00%	Interest rate of loan (annual)	10.00%

(Continued)

TABLE 13-5   Analysis of Different Types of Generators—Cont'd

Cost of Capital		Cost of Capital	
Amt. paid over life of loan	$9747	Amt. paid over life of loan	$12,745
Payments per year	*$1218/year*	*Payments per year*	*$1593/year*
TOTAL	**$1218/year**	**TOTAL**	**$1593/year**
Cost per kilowatt-hour	*$0.371 $/kWh*	*Cost per kilowatt-hour*	*$0.173 $/kWh*
kWh produced annually	*3285 kWh*	*kWh produced annually*	*9198 kWh*

Stirling Engine		Fuel Cell	
Fuel Cost		*Fuel Cost*	
Percent of time in operation	75.00%	Percent of time in operation	75.00%
Rated power output	1 kW	Power output	5 kW
Fuel conversion efficiency	25.00%	Efficiency	35.00%
Fuel consumption	4 MJ/kW	Fuel consumption	14.28571 MJ/kW
Fuel cost per energy	$0.00379/MJ	Fuel cost per energy	$0.05451/MJ
Annual fuel cost	*$358/year*	*Annual fuel cost*	*$18,418/year*

Cost of Capital		Cost of Capital	
Rated capacity	1 kW	Rated capacity	5 kW
Est. price per kW	$6000/kW	Est. price per kW	$5000/kW
Period of loan	8 years	Period of loan	8 years
Amount of loan	$6000	Amount of loan	$25,000
Interest rate of loan (annual)	10.00%	Interest rate of loan (annual)	10.00%
Amt. paid over life of loan	$8997	Amt. paid over life of loan	$37,489
Payments per year	*$1125/year*	*Payments per year*	*$4686/year*
TOTAL	**$1483/year**	*"***TOTAL**	**$23,104/year**
Cost per kilowatt-hour	*$0.226 $/kWh*	*Cost per kilowatt-hour*	*$0.703 $/kWh*
kWh produced annually	*6570 kWh*	*kWh produced annually*	*32,850 kWh*

Solar Panel		Gasifier	
Fuel Cost		*Fuel Cost*	
Average capacity factor	35.00%	Percent of time in operation	75.00%
Power output	1 kW	Rated power output	80 kW
Efficiency	60.00%	Fuel conversion efficiency	25.00%
Fuel consumption	0 MJ/kW	Fuel consumption	320 MJ/kW
Fuel cost per energy	$0.00000/MJ	Fuel cost per energy	$0.00473/MJ
Annual fuel cost	*$0/year*	*Annual fuel cost*	*$35,811/year*

(Continued)

TABLE 13-5 Analysis of Different Types of Generators—Cont'd

Cost of Capital		*Cost of Capital*	
Rated capacity	1 kW	Rated capacity	20 kW
Est. price per kW	$6000/kW	Est. price per kW	$1300/kW
Period of loan	8 years	Period of loan	8 years
Amount of loan	$6000	Amount of loan	$26,000
Interest rate of loan (annual)	10.00%	Interest rate of loan (annual)	10.00%
Amt. paid over life of loan	$8997	Amt. paid over life of loan	$38,988
Payments per year	*$1125/year*	*Payments per year*	*$4874/year*
TOTAL	**$1125/year**	**TOTAL**	**$40,685/year**
Cost per kilowatt-hour	*$0.367 $/kWh*	*Cost per kilowatt-hour*	*$0.077 $/kWh*
kWh produced annually	*3066 kWh*	*kWh produced annually*	*525,600 kW*

(Source: Iowa Wind Energy Institute. 2001. *Wind hybrid electricity applications.* Iowa Department of Natural Resources. Des Moines, IA.)

To optimize electrical energy production of a fuel cell, electrodes are thus usually made flat with a thin layer of solid or liquid electrolyte sandwiched between them. The type of electrolyte defines the type and name of the fuel cell.

The most common types of fuel cells are:

- Polymer electrolyte membrane fuel cells (PEMFCs), where the electrolyte is an ion exchange membrane
- Alkaline fuel cells (AFCs), where the electrolyte is an 80% concentrated solution of potassium hydroxide (KOH)
- Phosphoric acid fuel cells (PAFCs), where the electrolyte is 100% concentrated phosphoric acid
- Molten carbonate fuel cells (MCFCs), where the electrolyte is alkali carbonates that in the high operating temperatures of this fuel cell form molten salts
- SOFCs, where the electrolyte is a solid, nonporous metal oxide

Fuel Cell Efficiency

Efficiency of a system is defined as the energy produced divided by the energy input. In the case of fuel cells, the energy produced is electrical energy and the energy input is the chemical energy of the fuel. Efficiency is expressed as:

Electric energy produced / Heat energy provided

where the denominator is a function of heating or calorific value of the fuel.

Calorific value is either lower (LCV) or higher (HCV) calorific value, depending on whether the enthalpy of vaporization of water has been added. For hydrogen, HCV = −285.84 kJ/mole and LCV = −241.83 kJ/mole. The LCV is most commonly used, since it gives higher values of efficiency.

With combustion engines, efficiency is limited by the Carrot limit, namely $(TI\text{-}TZ)/TI$ where TI is the maximum temperature of the heat engine and TZ the lower (exit) temperature in degrees Kelvin. The efficiency limit of a heat engine increases with increasing maximum temperature.

The opposite happens for fuel cells, as can be observed in Figure 13-3. It can also be observed that, contrary to what is commonly believed, heat engines can display higher

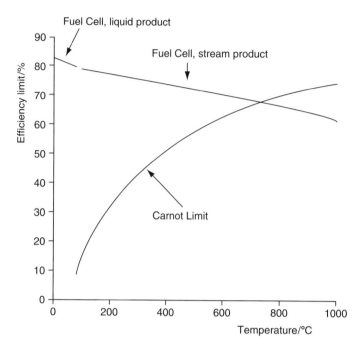

FIGURE 13-3   *Efficiency limits versus temperature plot for a fuel cell and a heat engine.* (Source: N. Lymberopoulos, N. 2005. Fuel cells and their application in bio-energy. *Project Technical Assistant Framework Contract [EESD Contract No: NNE5-PTA-2002-003/1].)*

efficiency limits than fuel cells (at high temperatures). However, at such temperatures, the heat available in a fuel cell can be used in combined heat and power (CHP) or in a bottoming cycle to produce more electricity through a gas turbine.

Reaction Rate

Like in any chemical reaction, the reaction of hydrogen at the anode of a fuel cell requires "activation energy" to start and then proceeds at a given rate of reaction. To reduce the activation energy and increase the rate of reaction, the application can use:

- Catalysts
- Higher temperatures
- Increased electrode area

Since reactions take place at the surface of the electrode, electrode area is vital. Performance of a fuel cell is quoted in terms of current per square centimeter (cm^2). The effective surface area of fuel cells is increased since electrodes have a microstructure that gives them two or three times higher surface areas than their planar area.

Bipolar Plates

The voltage across a single cell is typically 0.7 volts. To produce a useful voltage, many cells need to be connected in series, forming a "stack." The most common interconnection method is the bipolar plates. Such plates allow connections all over the cathode of one cell and the anode of the next. At the same time, the bipolar plates allow the feeding of the hydrogen-carrying gas to the anode and oxygen to the cathode, keeping the two gases tightly separated. This is why bipolar plates are sometimes referred to as *flow field* plates.

Other Fuel Cell Components (Balance of Plant)

The heart of a fuel cell is the stack, consisting of the previously discussed electrodes, electrolytes, and bipolar plates. The many other components are called *balance of plant* (BoP). These components depend on the type of fuel cell. They may take up more space than the fuel cell itself, for instance, in CHP applications.

The hydrogen-carrying fuel and the oxygen need to be circulated using pumps or blowers. Compressors could be used in some cases of larger fuel cells. Also, fuel cells produce DC power, so a DC/DC converter or a DC/AC inverter is probably required to feed power to the grid or to drive motors connected to the pumps or blowers. The processing of hydrogen could require complex hardware in case hydrogen needs to be purified or humidified, or in case pure hydrogen is not available, to reform and purify hydrogen-rich gases. Control valves, pressure regulators, and a controller are required as the start-up and shutdown of a fuel cell can be a complex procedure that the controller manages. A cooling system may be required, although in most cases the excess heat could be used for heating or preheating the fuel or to help release hydrogen in the case of hydrogen storage in metal hydride tanks.

Advantages of Fuel Cells

The advantages of fuel cells are:

- High efficiency: Fuel cells are capable of converting chemical energy of a fuel directly into electricity through an electrochemical reaction. One step is required compared to the three steps required in conventional installations that involve combustion for the production of thermal energy in the form of steam, expansion of steam through a steam turbine to produce mechanical work, and rotation of an electrical generator to produce electricity. Additionally, the efficiency of a fuel cell is independent of its size, which is not the case for today's combustion-based power equipment. This, however, does not hold for the BoP, where the load is proportionately higher for smaller sizes. Electrical efficiencies ranging between 35% and 55% can be achieved by fuel cells. If the thermal energy generated by the fuel cell is also converted to electricity through a turbine cycle, then the electrical efficiency can increase even up to 70%. In CHP mode, fuel cells can exhibit efficiencies up to 80%.
- Size flexibility: A single cell of a fuel cell produces approximately 1 volt of potential. Single cells can be stacked to provide the appropriate voltage for any application. Manufacturers produce fuel cells with a power of a few watts for mobile phones, a few tens of watts for portable computers, a few kilowatts for home systems, and a few megawatts for power-generation applications.
- High reliability: Fuel cells have no moving parts and thus require minimum maintenance and fewer shutdowns. However, BoP hardware can fail as it is often not designed specifically for fuel cells. The performance and reliability of some types of fuel cells are affected by the purity of the fuel, so purification equipment is important.
- Emissions: Water vapor is the only by-product if pure hydrogen is used as fuel. Even in the case of using fossil fuels like natural gas, there are no NO_x, carbon monoxide (CO), or hydrocarbon (HC) emissions. Carbon dioxide (CO_2) is emitted if fossil fuels are used or if the hydrogen comes from fossil fuels reforming or from fossil fuel-based electricity in electrolysers.
- Ease of installation: Since fuel cells have no moving parts, they produce minimal noise and vibrations. This, in combination with the limited emissions, makes installation simple and suitable for applications close or inside buildings, depending on local fuel regulations.

- Fuel flexibility: Given the appropriate fuel-treatment devices including reformers, purifiers, driers, and deoxidizers, various types of fuels can be used in fuel cells. There are types of fuel cells that are more tolerant to impurities or perform internal reforming, so fuel-treatment requirements are minimized. Fuel cells can be adapted to use hydrogen, methane, methanol, ethanol, landfill gas (LFG), digester gas, or oil.

Disadvantages of Fuel Cells

The disadvantages of fuel cells, on the other hand, are as follows:

- Cost: Fuel cells are expensive to build since demand has not reached commercial levels. Many devices are still custom built. Some fuel cells use expensive materials.
- Size and weight: Per unit of power produced, fuel cells have a larger footprint and weigh more than internal combustion engines. Their size and weight, however, are diminishing.
- Start-up times: The start-up times of fuel cells vary from a few minutes to some tens of minutes, which is a major drawback, especially for transport applications.
- Reliability: Even though reliability may eventually be higher than that of competing technologies, currently it is not.
- Fuel availability: For many fuel cell types and applications, the fuel is not readily available and related fueling infrastructure needs to be established. In case fossil fuels are used, their reforming produces some pollutants.

There are additionally some disadvantages that are related to specific types of fuel cells:

- PEMFCs require expensive catalysts and are susceptible to poisoning due to the low operating temperatures.
- AFCs require that the fuel and air streams are free from CO_2.
- PAFCs require platinum catalysts, run on low current and power, and have large size and weight relative to other fuel cell types.
- MCFCs and SOFCs that operate in high temperatures suffer from corrosion and breakdown of cell components.

Types of Fuel Cells

Proton Exchange Membrane Fuel Cells

PEMFCs, also referred to as solid polymer cells (SPs or SPFCs), were first developed in the 1960s for the NASA space program. Their electrolyte is an ion-conducting polymer membrane, onto the sides of which the anode and cathode are bonded (platinum catalyst), forming membrane electrode assemblies. PEMFCs operate at temperatures around 80°C, but efforts are made to increase this temperature to more than 150°C that will lead to a more tolerant cell to impurities.

Pure hydrogen is the ideal fuel for PEMFCs whereupon they can display a very high efficiency of 50%; however, due to the fact that they operate at low temperatures, the heat produced cannot be exploited for fuel reforming in case pure hydrogen is not available.

In such cases the efficiency could fall to below 40% (some of the fuel would be consumed to provide the heat for reforming). CO can poison the fuel cell catalyst and therefore additional hardware is needed to limit CO to below 50 parts per million (ppm).

PEMFCs have a high power density, meaning low weight, cost, and volume. This has resulted from increases in current density that can be as high as $1 A/cm^2$. Their low-temperature operation allows for shorter start-up times and better load-following capabilities.

PEMFCs are seen as the fuel cell of choice for transport applications but can also be applied in portable and stationary applications, including CHP.

PEMFCs can be applied to a wide range of power applications, from a few watts for mobile phones, to a few kilowatts for stranded generating sets, to a few tens of kilowatts for cars and buses, to a few hundreds of kilowatts for industrial CHP systems. All these systems would utilize the same type of electrolyte, electrode structure, and catalyst, but would differ in the water management, the method of cooling, the bipolar plate design, operating pressure, and the reactants used.

Alkaline Fuel Cells

Bacon demonstrated the viable operation of AFCs in the 1940s and 1950s. AFCs were used in the Apollo missions to the moon. AFCs are high-performance fuel cells due to the rate at which chemical reactions take place in the cell. They are also very efficient, with efficiencies of 60% in space applications.

The electrolyte of AFCs consists of a potassium solution (35 wt% KOH) for 120°C operating temperature that is more concentrated (85 wt% KOH) for an operating temperature of 260°C. The fact that the electrolyte is a corrosive liquid is considered one of the disadvantages of AFCs.

Another disadvantage is that it is easily poisoned by CO_2. Even the small amount of CO_2 in air can affect the cell's operation, making it necessary to purify both the hydrogen and oxygen used in the cell. This purification process is costly. Susceptibility to poisoning also affects the cell's lifetime, further adding to cost. Cost is less of a factor for remote locations such as space or under the sea. However, to effectively compete in most mainstream commercial markets, these fuel cells will have to become more cost effective.

AFC stacks have been shown to maintain sufficiently stable operation for more than 8000 operating hours. To be economically viable in large-scale utility applications, these fuel cells need to reach operating times exceeding 40,000 hours. This is possibly the most significant obstacle in commercializing this fuel cell technology.

Phosphoric Acid Fuel Cells

Operating fuel cells at higher temperatures to PEMFs or AFCs has some advantages, including:

- Faster electrochemical reactions due to lower activation losses
- Higher tolerance to contaminants, meaning reduced need for noble metals
- The high temperature allows for the reforming of the fuel used and the "extraction" of hydrogen
- Similarly, the heat available at the exit of the fuel cell can be used in applications requiring good-quality heat (i.e., CHP) or can be used to drive a gas turbine in order to produce more electrical power in fuel cell and turbine hybrid systems

PAFCs operate at medium temperatures (up to 200°C) and are the most well developed of the "hot" fuel cells. Many installed 200-kW PAFC CHP systems were built by United Technologies ONSI (formerly International Fuel Cells).

Some of these units have been run on biogas from sewage treatment plants. Since PAFCs operate at medium temperatures, they require noble metal catalysts made of platinum and, like PEMFCs, are poisoned by CO, meaning that a complex fuel-processing system is required.

The electrolyte is concentrated phosphoric acid (H_3PO_4). The electrodes used are of the gas diffusion type like for proton exchange membrane fuel cells (PEMFCs) consisting of Pt supported on carbon, a solution that has allowed considerable reduction in Pt loading.

The stack consists of a repeated arrangement of a bipolar plate, the anode, the electrolyte, and the cathode, like in a PEMFC. Provision must be made for cooling of the cells using air or liquids, the latter requiring less space between cells.

Some indicative performance figures of the most advanced cells produced by International Fuel Cells were 0.323 W/cm^2 in cells operating at 645 mA/cm^2 and 0.66 volts/cell. Degradation rates of 4 mV/1000 hours of operation have been achieved, but these have been surpassed by Mitsubishi Electric with 2 mV/1000 hours for 10,000 hours of operation at 200 to 250 mA/cm^2.

A number of PAFCs (more than 65 MW) have been installed and operated for many years and have shown very good reliability of the stack and power quality. This means that such systems have been chosen in "premium power" applications such as banks, hospitals, or computing facilities. Most plants are just 200 kW, but some plants go up to 5 MW. A specific unit built by International Fuel Cells and Toshiba has a power of 11 MW.

Molten Carbonate Fuel Cells

MCFCs are high-temperature fuel cells that use a molten carbonate salt mixture electrolyte suspended in a porous, chemically inert ceramic LiA1O2 matrix. Since they operate at extremely high temperatures of 650°C and above, nonprecious metals can be used as catalysts at the anode and cathode, reducing costs. MCFCs are currently being developed for natural gas–fueled stationary application of few megawatts rating for power plants for electrical utility and industrial applications.

Molten carbonate fuel cells can reach efficiencies approaching 60%, considerably higher than the 40% efficiencies of a phosphoric acid fuel cell plant. When the waste heat can be exploited, overall fuel efficiencies can be as high as 85%.

Unlike the fuel cells operating at low or medium temperatures, MCFCs do not require an external reformer to convert more energy-dense fuels to hydrogen. Due to the high temperatures at which they operate, these fuels are converted to hydrogen within the fuel cell itself by a process called *internal reforming,* which also reduces cost.

MCFCs are not prone to CO or CO_2 "poisoning"—they can even use CO_x as fuel—making them more attractive for fueling with gases produced from coal. This makes them particularly suitable for using LFG or anaerobic digester gas (ADG) produced at wastewater treatment plants where CO_2 can be as much as 40% of the produced gas.

Although MCFCs are more resistant to impurities than other fuel cell types, scientists are looking for ways to make them resistant enough to impurities from coal, such as sulfur and particulates.

The primary disadvantage of current MCFC technology is durability. The high temperatures at which these cells operate and the corrosive electrolyte used accelerate component breakdown and corrosion, decreasing cell life. Scientists are currently exploring corrosion-resistant materials for components as well as fuel cell designs that increase cell life without decreasing performance.

As fuel cells are a relatively new field, good illustrations are not as easy to find as they would be with a more mature technology such as gas turbines. The illustrations that are available can be found on original equipment manufacturers' Web sites. One manufacturer's Web site to check is http://www.fce.com/.

Solid Oxide Fuel Cells

SOFCs use a hard ceramic compound as the electrolyte. Since the electrolyte is a solid, the cells do not have to be constructed in the plate-like configuration typical of other fuel cell types and cylindrical cells are the norm. SOFCs are around 50% to 60% efficient at converting fuel to electricity. In CHP applications, overall fuel use efficiencies could top 80%.

SOFCs operate at very high temperatures (around 1000°C) that remove the need for precious-metal catalyst, thereby reducing cost. It also allows SOFCs to reform fuels internally. This allows the use of a variety of fuels and reduces the cost associated with adding a reformer to the system.

SOFCs are also the most sulfur-resistant fuel cell type; they can tolerate several orders of magnitude more sulphur than other cell types. In addition, they are not poisoned by CO, which can even be used as fuel. This allows SOFCs to use gases made from coal.

High-temperature operation, however, has disadvantages. Long start-up times are encountered (making them unsuitable for transport applications) plus significant thermal shielding is required to retain heat and protect personnel.

The high operating temperatures also demand stringent material durability requirements. The development of low-cost materials with high durability at cell operating temperatures is the key technical challenge facing this technology.

Areas of current research are development of lower-temperature SOFCs operating at or below 800°C that have fewer durability problems and cost less. Lower-temperature SOFCs produce less electrical power, however, and stack materials that will function in this lower temperature range have not been identified.

SOFC-based home CHP units have been developed in Europe, and beta models are being tested in the context of demonstration campaigns. SOFCs are ideal to be hybridized with gas turbines thanks to the high operating temperatures. Illustrations, application examples, and further information are provided on Web sites maintained by original equipment manufacturers, for instance, Siemens (Figure 13-4).

Direct Methanol Fuel Cells

The direct methanol fuel cell (DMFC) converts methanol and oxygen electrochemically into electrical power, heat, CO_2, and water. This type of fuel cell is in contrast with one

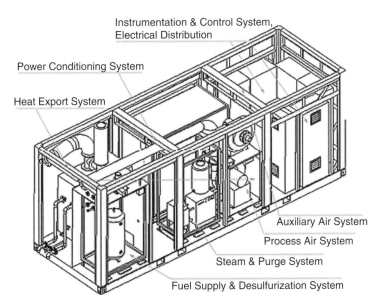

FIGURE 13-4 *The Siemens Westinghouse 220-kW SOFC-gas turbine hybrid. The SFC-200 will be the building block for systems up to 500 kW. (Source: Siemens. 2007. SOFC Product Commercialization. Available at: http://www.powergeneration.siemens.com/en/fuelcells/commercialization/index.cfm.)*

that uses hydrogen that has been produced from the reforming of methanol, which can be termed an *indirect methanol fuel cell*. A DMFC works in the same way as a PEMFC with the difference that, at the anode (negative electrode), the methanol is first split into hydrogen and CO_2 before the same catalyst splits the hydrogen into protons and electrons. The reaction at the anode is shown below:

$$CH_3OH + H_2O \rightarrow 6H^- + 6e^- + CO_2$$

Methanol needs to be mixed with water. The reactions involved are rather slow and require a special platinum/ruthenium catalyst on a carbon substrate to improve reaction rates. Similarly, the cathode reaction is catalyzed by platinum particles on a carbon substrate. The major drawback of DMFCs is their low efficiency that is quoted as 4% due to the slow rates of reaction. For improved rates of reaction through the catalysts mentioned previously, electrical efficiencies of the order of 30% have been quoted (Figure 13-5).

Direct Ethanol Fuel Cells

The previously described DMFC technology has been investigated for a number of years, leading to commercial products. The direct use of ethanol (or other small organic compounds such as formic acid or formaldehyde) in PEMFCs is still in an early investigational stage, with experiments being performed with single-cell fuel cells. Such fuels are being investigated in order to identify fuels that are largely available or renewable, inexpensive, nontoxic, and with a high reactivity at low temperatures.

The main problem of direct ethanol (sometimes referred to as *direct alcohol*) fuel cells is their low efficiency, resulting from the poor performance of the electrocatalysts and the severe fuel crossover from anode to cathode. Research to increase the reaction rate at the anode that will increase efficiency and reduce fuel permeation (the latter caused by reduced fuel concentration at the anode as the fuel is consumed more exhaustively) is in progress.

Fuel Cells Research

A number of countries are undertaking major efforts for the development of fuel cells. Japan has arguably the biggest national program that in a period of 20 years has spanned high- and low-temperature fuel cells, for stationary or transport applications. Current research focus is on PEMFCs for embedded generation in buildings (domestic and office).

FIGURE 13-5   *A 700W DMFC of IRD Fuel Cells A/S. Available at: http:// [IRD Fuel Cells web site]*

Canada has a similarly long and important involvement in fuel cell development and has the industry to prove its success. Effort has been based on low-temperature fuel cells for stationary, transport, and stand-alone applications.

Canada has indeed been a pioneer in the development of PEMFCs, with Ballard being one of the most successful commercial fuel developers. Besides developing and commercializing fuel cells, the national efforts focus on the development of standards for hydrogen and fuel cell technologies in view of their wider application.

In the United States, research and development efforts are led by the DOE and focus on the development of reliable, low-cost, high-performance fuel cell system components for transportation and stationary applications. The available budget is about $40 million/year, some of which is allocated towards the Vision 21 projects for the development of clean central station generation technologies and the rest to embedded generation technologies.

A major U.S. activity is the Solid State Energy Conversion Alliance, a partnership between the DOE, the National Laboratories, and industry to develop and demonstrate planar SOFCs for embedded generation. The DOE also develops fuel cell technologies with an emphasis on the PEMFC for both stationary and transportation applications.

PEMFCs for transport applications are being developed in the context of the FreedomCAR partnership between the DOE and the major car manufacturers of the United States. The DOE also has the responsibility for developing PEMFCs for portable and distributed generation applications as well as the technologies required for the hydrogen energy infrastructure that is important in the long term for large-scale use of PEMFCs.

Europe has been a leader in hydrogen technologies and especially in hydrogen production technologies through pressurized electrolysis. In the last 20 years, its research agenda in the field expanded to include fuel cells.

Considerable research and development on fuel cells has been conducted in Germany in the last 15 years, Germany being a leader in hydrogen technologies. Efforts have concentrated on new materials, improved components, and system integration.

The national fuel cell program has received strong national and regional funding and has made notable progress in a number of key areas including development of high-temperature fuel cells (e.g., MCFC, SOFC) for stationary applications, including the 200-kW Hot Module MCFC of MTU and the 1-kW Hexis SOFC of Sulzer for home applications. Public funding for fuel cell activities amounts to 10 million euros annually and an equal amount is provided by the private sector.

In Italy, established heavy industries as well as young "small" players have been involved in research and development activities in areas of fuel cell development since the early 1980s. Currently, they are focused on the development and demonstration of PEMFCs for stationary and automotive applications and MCFCs for on-site and distributed generation. Research activities on materials and components for SOFCs are also being carried out. A national program of 30-million-euro annual budget supports research and development efforts in the field.

BIOFUELS FOR FUEL CELLS

Raw biomass can be used directly as fuel; however, in many cases it is processed prior to usage in order to be converted to a form more suitable to various energy technologies. The various biomass conversion technologies are briefly outlined in this section along with the compatibility of the produced biofuels to fuel cells.

The average cost of various fuels is compared with that of gaseous biofuels in Table 13-6.

TABLE 13-6   Average Cost of Gaseous Fuels

Fuel Cost	(Euro per MMBTU)
Reformed H_2	17
Natural gas	6
Fuel gas from biomass gasification (BG)	>40
Biogas from anaerobic digestion of biomass (ADG)	1.2
Landfill gas (LFG)	1.6 to 2.5
Ethanol	10 (sugar-based)
	12–15 (cellulosic)

Fuel Gas From Biomass Gasification

Biomass gasification is a well-established technology for the production of combustible gas from various biomass feedstocks including wood, wood chips, forest waste, and agricultural and even municipal waste. Gasification is a two-step endothermic process during which a solid fuel is converted into a combustible gas.

First, components of the fuel are pyrolized, whereupon volatile components are vaporized at temperatures around 600°C. The resulting vapor contains HC, hydrogen, CO, CO_2, tar, and water vapor. In the second step, char (that had not vaporized in the first step) is gasified through reactions with oxygen, steam, and hydrogen. The end product is a fuel gas that can be upgraded to synthesis gas after purification and removal of pollutants and particulates. Some of the unburned char is combusted to provide heat to the endothermic reactions.

In case these gasification systems are integrated in CHP plants that use conventional gas-fired engines or turbines, then overall efficiencies of 25% to 35% can be achieved. For larger-scale systems like an integrated gasification combined cycle (IGCC), efficiencies of up to 40% to 45% can be achieved. Such values can be achieved with fuel cells even for smaller scales on the order of a few kilowatts.

The composition of the fuel gas produced depends on the original composition of the feedstock as well as the process parameters. The gas composition is usually hydrogen (30% to 40%), CO (20% to 30%), CH_4 (10% to 15%), water (6%), nitrogen (1%). However, contaminants present in this fuel gas need to be removed for use in fuel cells.

The purification process varies according to the gasification technology and the type of fuel cell. Diluents and contaminants like tar, particulates, sulfur, and alkali metals need to be removed. Reforming, where CH_4 is converted to CO_2 and H_2 and cleanup occurs, is one method. That can be done as follows:

- Leaching: Removal of soluble organics and alkali compounds by "washing" biomass prior to gasification
- Activated carbon: Adsorption of contaminants after gasification
- Cold sulfur removal: Use of a fluidized-bed tar cracker followed by scrubbing
- Hot gas cleanup: Use of a cracking process to convert tars and unreacted char to hydrogen sulfide (H_xS), CO, and light HC

These methods are well established; however, their application at the scale of small (a few kilowatts) fuel cells is still to be proven from a practical and economic point of view. Indeed, there is no consensus in scientific literature as to whether a fuel cell of a gas turbine would be more suitable and tolerant to use the gas produced from biomass gasification, especially in the case of CHP applications.

Considering the various types of fuel cells, high-temperature fuel cells like SOFC and MCFC are considered as more appropriate to use than syngas since, at the temperatures they operate, CO is a fuel rather than a contaminant. Tars and sulfur need to be removed from the gas stream. Low-temperature fuel cells on the other hand do not integrate well with biomass gasification due to their lower tolerance to contaminants, meaning that the purification sections would be more complex and costly.

Biogas

The anaerobic digestion of biomass produces methane. Such facilities are commonly used by agricultural facilities, agro-food industries, animal farms, and wastewater treatment plants as a waste treatment methodology that promotes nutrient recycling and odor control. Bacteria are used to promote biological reactions that convert approximately 40% of the solid input into a methane-rich gas.

Biogas from anaerobic digestion (ADG) typically contains 55% to 65% methane, 30% to 40% CO_2, 1% to 10% nitrogen, small amounts of oxygen, and traces of substances that can poison fuel cells like hydrogen sulfide, halogen compounds, and non-methane organic compounds. There are two basic types of digesters:

- *Batch type digesters* break up one batch of material at a time. They are loaded with waste and then capped. Common types are the covered lagoon and the complete mix digesters. Waste decomposition produces heat, but adding heat will accelerate the process. Mixing is used to keep the solids (2% to 10%) in suspension.
- In *continuous digesters*, waste is fed continuously or at regular intervals. A common type is the plug flow digester. Wastes with 11% to 13% solids can be processed.

In addition to the feeding mode of operation of a digester, there are two basic principles of digester operation: (1) the liquid phase fermentation, in which the solid phase is rather low and usually in the range of 4 wt% to 10 wt%, and (2) the solid state fermentation, where the solid phase is in the range of 30 wt% to 35 wt%. The latter type can be considered as an intensification process since it has significantly higher biogas productivity per day per unit reactor volume (by a factor of 5 to 7).

Temperature is the single most significant factor in both digestion processes. Bacteria are most productive in the range 36°C to 54°C, and their operation can be either mesophilic (30°C to 38°C) or thermophilic (50°C to 70°C). This shows that there is potential for CHP applications at such installations, exploiting the ADG produced, which is either released or flared. Some internal combustion engines have been installed, but these are rather noisy and with considerable emissions, so fuel cells could prove an interesting alternative.

ADG can be used in distributed power and heat applications, like for the case of fuel cells mentioned below. However, it can also be upgraded and injected to the natural gas network to be used in more conventional applications.

The process of upgrading before injection to the natural gas grid involves CO_2 and hydrogen sulfide removal so as to increase the methane concentration by about 30%, increase the heating value by 7.5 to 11 kWh/normal cubic meters (Nm^3), reduce CO_2 by 30%, and reduce the H_2S from 500 ppm down to 3 ppm. Such a project in Pucking, Austria, where 10 Nm^3/hr of ADG is processed and about 6 Nm^3/hr is fed to the natural gas grid, will cost 1.1 million euro.

In Tilburg, The Netherlands, biogas is produced from source-separated waste and garden waste and is purified to natural gas quality and subsequently fed into a natural gas pipeline of the local network. Another attractive biogas application is in the transport sector where biogas is upgraded to natural gas quality and is compressed for use in buses as transport fuel, as it is in the cities of Lille and Stockholm.

For a fuel cell application running on wastewater treatment plant biogas, the contaminant limits are shown in Table 13-7, as presented in a study by the U.S. Environmental Protection Agency.

To remove the various contaminants for the above-mentioned case, a filter was installed prior to the pretreatment facilities to remove solids, liquids, and bacteria. The pretreatment facilities consisted of a non-regenerable desulfurizer bed in conjunction with a coal bed and were used to remove hydrogen sulfide with an efficiency of 98%.

Halide levels were sufficiently low for them to be handled in a halogen guard bed available at the fuel cell processor itself. The previous treatment was considered as adequate to offer a 5-year life span to the fuel cell's catalyst.

Since fuel cells suitable for such applications are commonly manufactured to operate on natural gas, some alterations need to be made to allow for the handling of larger quantities of lower calorific value gases, since ADG contains around 35% to 45% diluents (mostly CO_2). These modifications included resizing pipework and valves to allow for increased flow capacity and pressure. The actual operation showed lower values of efficiency than the respective fuel cell running on natural gas. Since the price of fuel is much lower than that of natural gas, this can compensate for loss of efficiency. A more significant drawback, however, is the fact that the methane content of biogas can vary between 10% and −10%. As a result, the fuel cell needs to be operated at power outputs lower than the nominal capacity in order to avoid shutdown related to any decrease in CH_4 content.

Biogas is compatible with PAFCs, MCFCs, and SOFCs. Most demonstration projects have used PAFCs but some demonstrators are planned with MCFCs.

Landfill Gas

Methane is generated in landfills by the natural degradation of municipal solid waste by anaerobic micro-organisms. A pipe grid system can be used to collect this gas and lead it to a flare or to a CHP system, preventing the release of methane to the atmosphere (20 times more potent greenhouse gas than CO_2) as well as controlling odor.

TABLE 13-7   List of Biogas Contaminants That Need to Be Removed Prior to Use in Fuel Cells

Biogas Contaminant	*PAFC Requirements*	*Issue/Concern*
Hydrogen sulfide (H_xS)	<4 ppmv	Poison to fuel processor reforming catalyst
Halogens (F, Cl, Br)	<4 ppmv	Corrosion of fuel processor components
Non-methane organic compounds	<0.5% olefins	Poison to fuel processor shift catalysts
O_2	<4%	Over-temperature of fuel processor beds due to excessive oxidation
NH_3	<1 ppmv	Fuel cell stack performance
H_2O	Remove moisture and condensate	Damage to fuel control valves Transport of bacterial phosphates
Bacteria and solids	Remove all bacteria and solids	Fouling of fuel processor piping and beds

PAFC = phosphoric acid fuel cells; ppmv = parts per million by volume.
(Source: Spiegel, 2000.)

LFG is composed of 50% to 60% methane, 40% CO_2, and small quantities of nitrogen, hydrogen, oxygen, and H_xS. Many other compounds are found as traces, like alkalines, aromatics, chlorocarbons, HC, oxygenated compounds, and sulfur compounds. Due to the complexity of its composition, four or five separate steps are required for the pretreatment of LFG before it can be used in fuel cells.

The gas entering the fuel cells must be essentially clean of contaminants like sulfur and halogens and consist primarily of methane, nitrogen, oxygen, and CO_2. Besides the obvious factors affecting the gas pretreatment steps that are the composition of the LFG and the requirements of the fuel cell, a third factor is that the composition of the LFG can vary periodically in the context of seasons or years.

The treatment steps for cleaning LFG for subsequent use in fuel cells are as follows:

- H_xS removal at ambient temperature
- Cooling
- Condensation
- Drying
- Additional cooling
- HC removal
- Filtration

However, a simple desulfurization step proved effective in the experimental running of a small SOFC (3-cm^2 single cell) with LFG in a study in the year 2000. After running the fuel cell for 6 hours, its power was reduced to only 70%. Further improvements were achieved by mixing the LFG with air in order to augment the inherent CO_2 reforming with partial oxidation.

Most of the projects that have demonstrated the use of LFG with fuel cells have relied on PAFCs, which were the first fuel cells that were commercially available and at significant capacities. PAFCs will most probably be the fuel cell of choice in the near term but will be replaced in the long term by SOFCs and MCFCs that require fewer pretreatment steps, once these units become commercially available. PEMFCs are not that suitable for using LFG.

Ethanol

Ethanol is produced in large quantities around the globe and is used as a petrol additive or replacement in the transport sector. In the first case, conventional internal combustion engines can be used provided the amount of ethanol does not exceed 5%, while in the second case, the internal combustion engine must have separate specifications. Ethanol can be used well in fuel cells. Ethanol can be produced in two ways:

- Through technologies that convert *starch or sugar-based* raw material.
- Through technologies that convert *cellulosic biomass* into ethanol. In this case, sugars must be formed from the cellulosic material, which subsequently can be fermented and distilled into ethanol.

Due to the high purity of the resulting ethanol, there is minimal pretreatment that needs to be done to the fuel for its use in fuel cells, compared to other biofuels. A reforming process is certainly necessary in order to produce hydrogen, which can be either steam reforming or partial oxidation.

For stationary applications, steam reforming is more suitable for the case of dilute ethanol and water mixtures (55% ethanol and 45% water by volume, resulting from partial distillation) while partial oxidation is preferable for more pure ethanol mixtures.

Ethanol is considered as a better fuel than natural gas in terms of energy density, cell voltages, and electrical power density (W/cm^2) that a fuel cell would provide. However, ethanol is more expensive on a cost-per-kilowatt-hour basis compared to natural gas.

High-temperature fuel cells like MCFCs and SOFCs are capable of internally reforming ethanol. For lower-temperature fuel cells, an external reformer is required that will operate at high enough temperatures (higher than those of the fuel cell) in order to turn ethanol into a viable fuel that will not cause poisoning of the fuel cells.

Ethanol can well be used in fuel cells for transport applications. Indeed, it has a number of advantages in terms of safety and storage requirements in comparison to pure hydrogen or gaseous biofuels. However, its cost is again the main drawback for its application. It is expected that this cost could fall to $0.15/liter by 2010. The present high cost means that there are today very few demonstration facilities.

Besides the previous use of ethanol in fuel cells that involves a reforming step, there do exist fuel cells that can use ethanol directly, commonly referred to as *direct ethanol fuel cells*.

Pyrolysis Oil

Pyrolysis oil is produced through a pyrolytic process wherein biomass material (commonly forest or agricultural waste) is rapidly heated to 500°C in the absence of oxygen and then vaporized and condensated to a liquid oil. Due to the complexity of pyrolysis oil in terms of the different compounds encountered, little research work has been done with respect to its use in fuel cells. However, some studies have been made that consider that reforming of pyrolysis oil is technically feasible. The potential "cooking" of the catalysts could prove to be a major obstacle, however. Another approach currently investigated is catalytic reforming that would convert pyrolysis oil into gases, mostly hydrogen and CO (at a 2:1 ratio), some methane (2% to 4%), and CO_2 (15%).

Pyrolysis oil is expected to be able to achieve lower prices than ethanol; however, the complexity of the reforming process means that a higher-temperature reformer would be required, meaning in turn that more energy would be required in that step. Pyrolysis oil is currently being demonstrated as fuel in more conventional combustion engines like gas turbines or internal combustion engines. Its use, compared to conventional fuels, would result in a carbon-neutral cycle and lower SO_x emissions plus 2% NO_x emissions compared to a diesel engine.

FUEL REFORMING

Reforming is the generic term used for converting HC into hydrogen and CO_2. There are three basic reforming techniques: steam reforming (endothermic), partial-oxidation (exothermic), and autothermal reforming (combination of the previous two; close to thermal equilibrium). The reforming process consumes between 20% to 30% of the energy contained in the fuel to be reformed.

STATIONARY AND TRANSPORT APPLICATIONS OF FUEL CELLS

Fuel cells are a suitable energy-conversion device for both stationary and transport applications. Low-temperature fuel cells seem at present more suitable for the transport sector due to their lower operating temperatures that mean short start-up times. On the other hand, PEMFCs require hydrogen of high purity, meaning that it would be preferable for a vehicle to carry pure hydrogen rather than a hydrogen-rich fuel that would need a heavy and space-occupying reformer and purifying unit.

High-temperature fuel cells are more suitable for stationary applications, especially for the case where the heat generated can be utilized (CHP). Their long start-up times are not a problem for continuously running applications while their capability to tolerate contaminants or perform internal reforming is an advantage in case an existing fuel distribution infrastructure exists.

Biofuels can be used in such stationary and transport applications, as described below.

Stationary Applications—Distributed Generation

Stationary applications of fuel cells for the production of electricity and heat fall under the term *distributed generation*. A number of parameters that include the liberalization of the energy market, environmental concerns, and security of supply have increased the interest in this concept in contrast to centralized generation. This interest has been boosted by the spreading of the natural gas networks and by the fact that power transmission losses are avoided if power is produced locally. The possibility to exploit on-site the heat produced in CHP installations further increases the benefits of the distributed generation concept. Suitable technologies have been developed for distributed generation CHP applications including the following:

- Advanced internal combustion engine based CHP systems
- Microturbines
- Stirling engines
- Fuel cells

The strengths of distributed generation can be summed up as follows:

- Low investment costs
- High efficiency, up to 80% for CHP
- Short times for installation
- Installation close to load, avoiding transmission losses and power line refurbishment or extension
- Low emissions
- Capability to utilize a variety of fuels

The weaknesses of distributed generation are as follows:

- Relatively high cost of kilowatt-hours, depending on fuel used
- Need for attention to electrical issues like control of voltage, frequency, and reactive power
- Non-technical issues for connecting to the electricity grid
- Installation cost can be high in the case of existing buildings that need a retrofit

Densely populated areas (that in Europe accommodate 80% of the population) offer major opportunities for the development of such distributed small-scale cogeneration schemes since the demand for power and heat concentrates there. The ideal fuel for these applications would be city gas or natural gas. However, rural areas also constitute an interesting market. In rural areas, the benefits of reduced transmission losses become more obvious, while the investment in electricity transportation infrastructure is avoided. Such systems would allow commercial activities to take place, contributing to rural development programs.

Fuel cells would offer a number of potential advantages compared to other distributed-generation technologies, including improved electrical efficiency, low noise, minimal emissions, and small number of moving parts (found only in the peripheral systems). Table 13-8 compares typical emission values for fuel cells and other distributed-generation technologies utilizing natural gas for power production.

Fuel cells, however, have some drawbacks that still need to be addressed. These are the high cost of fuel cells and of the upstream fuel pretreatment and purification hardware and the still limited operating life. The fact that the existing electricity distribution networks are unsuitable for accommodating large numbers of small generation plants will not help in the penetration of distributed-generation technologies, including fuel cells.

Transport Applications

Biofuels have a role to play in the European Commission's Green Paper on security of energy supply and in the White Paper on a common transport policy in which, for the year 2020, a target of 20% of alternative fuels in road transport has been set. The most important of these alternative fuels are as follows:

- Natural gas
- Biofuels
- Hydrogen

According to European Commission directive 2003/30, biofuels are to cover 2% of the transport fuel market by the year 2005 and 5.75% by the year 2010.

All these power trains can be combined with batteries for intermediate power storage and regenerative braking capability, leading to hybrid configurations.

FUEL CELL CASE STUDIES

There are very few installations around the world with fuel cells operating on biomass or waste-derived fuels. On the other hand, a number of projects run laboratory single-cell fuel cells with biogas from landfill sites or wastewater treatment plants in order to optimize the fuel cell components or the pretreatment methods. Commercial-size applications, however, are very rare indeed. Only one such installation was located in Europe, while some few installations exist in the United States and Japan. These are presented below.

TABLE 13-8   Typical Emission Values for Fuel Cells and Other Distributed Generation Technologies

Technology	Rating	NO_x	CO	HC
Microturbines	30–100 kW	9–25 ppm	25–200 ppm	9–25 ppm
Gas turbines	0.8–10 MW	6–140 ppm	1–460 ppm	6–560 ppm
Internal combustion engines	35 kW	30–450 ppm	240–380 ppm	—
Internal combustion engines	0.17–0.5 MW	30–3200 ppm	320–830 ppm	2750 ppm
PAFC	200 kW	1 ppm	2 ppm	—

NO_x = nitrogen oxide; CO = carbon monoxide; HC = hydrocarbon.

Koln Rodenkirchen Waste Water Treatment Plant, Germany

A 200-kW PAFC was installed at this wastewater treatment plant that treats water from 70,000 people and produces 1500 to 2000 Nm^3 of ADG per day. The unit was installed by RWE Fuel Cells GmbH in 2000 and was the first such unit in Europe. A gas-processing unit was commissioned the previous year.

The reasons that this particular type of fuel cell was chosen were as follows:

- The fuel cell ADG consumption (90 Nm^3/hr) matched the gas production capacity of the site.
- The fuel cell exhibited excellent part load characteristics between 50% and 100% load.
- The usable heat and temperature matched the requirements of the fermenters.
- It was the only commercially available fuel cell of this size in 1998.
- It had already been tested with ADG.

The project's duration was 5 years, with a load time of over 29,500 hours and a longest run of 5000 hours. The availability of the fuel cell was 80% and of the fuel cell system 70%. More than 1.8 MNm^3 of ADG were consumed at an average electrical efficiency of 37%, producing more than 4 MkWh.

Gas processing was done in two stages:

- First stage: gas freezing
- Second stage: activated carbon

The parasitic load of the plant was 3.5 kW and required service every 6 months (activated carbon).

The conclusions drawn after 5 years of operation were as follows:

- The site preparation and fuel cell installation was a straightforward operation.
- The two-stage gas processing produced a gas clean enough to be used by fuel cells.
- The fuel cell was capable to run solely and continuously on ADG.
- The fuel cell was considered to outperform the common gas engines.
- The operation of the fuel cell could be handled by the wastewater treatment plant staff, but monitoring and support by qualified engineers was necessary.
- Getting a permit was difficult due to the fact that the technology was imported.
- The reliability was lower than expected.
- A number of shutdowns occurred due to factors external to the fuel cell.
- Automatic start-up procedures did not work on ADG.
- Gas should be processed immediately after the fermenter in order to protect all hardware downstream.

Penrose Power Station, Sun Valley, California, U.S.

A 200-kW PAFC was installed at the Sun Valley, CA, landfill site in 1994 and was operated until 1995 with LFG. The unit was able to produce a peak power of 137 kW and a steady power output of 120 kW with an efficiency of 36.5%. Tests lasted for a total of 700 hours with an availability of 98.5%.

Groton Landfill, Connecticut, U.S.

The PAFC unit of the Penrose test site was moved to the Groton Landfill, installed in 1996, and tested until 1997. The respective values measured were peak power of 165 kW and a steady power output of 140 kW with an efficiency of 38%. Tests lasted for a total of 3300 hours with an availability of 96.5%.

Portland, Oregon, U.S.

A 200-kW PAFC unit of UTC was installed at the wastewater treatment plant at Portland and operated on anaerobic digester gas (ADG). The unit was installed in 1999 and was operated for 13,000 hours with an average power of 136 kW, an availability of 82%, and an electrical efficiency of 38%.

Calabasas, California, U.S.

At another wastewater treatment plant in Calabasas, CA, fuel cell PAFC technology was applied. Two 200-kW fuel cells were installed and were operated for 6850 hours. Availability was 91.2% and almost 1000 MWh was generated.

King County, Washington, U.S.

A 1-MW MCFC unit of FuelCell Energy Inc. was installed at a cost of $22 million at the wastewater treatment plant of King County, WA. The site is currently processing wastewater from about 1.4 million people, producing biogas that is scrubbed and is provided to the local natural gas network. The quantities produced can generate 4 MWe (megawatt electrical). The fuel cell installed (the single largest unit in the world) will use the biogas produced more efficiently than the current gas scrubber-network combination.

Kirin Brewery, Tokyo, Japan

A 250-kW MCFC of FuelCell Energy Inc. was installed at the Kirin Brewery near Tokyo. The fuel cell is operated in cogeneration mode, using a methane-rich digester gas produced from the effluent from the brewery process. The thermal output of the fuel cell is used by the anaerobic digester, which treats the brewery effluent.

THE HYDROGEN ENERGY VECTOR

Environmental concerns and security of supply issues support the transition from a fossil fuel–based society to a hydrogen society in order to meet our ever-increasing energy needs in a sustainable manner. Historical trends prove that humanity, once in the industrial age, tends to use fuels whose carbon content keeps diminishing, with the hydrogen content increasing, rendering hydrogen as the "ultimate" fuel. The combination of hydrogen, biofuels, electricity, and fuel cells gives a promise for a sustainable energy future for Europe and the world.

Hydrogen Properties

Hydrogen is the most abundant element in nature but can be found only in compounds due to its high reactivity (e.g., water, HC), which on the other hand makes hydrogen such an interesting fuel, suitable for many combustion applications (Table 13-9). Hydrogen

TABLE 13-9   Comparison of Energy Properties of Various Energy Carriers

Energy Carrier	H_x (220 bar)	Methane (NG)	LPG	Methanol	Petrol	Lead batteries
Energy density per weight (kWh/kg)1	33.3	13.9	12.9	5.6	12.7	0.03
Energy density per volume (kWh/lt)	0.53	2.6	7.5	4.4	8.7	0.09

LPG = liquified petroleum gas.

can be produced from a variety of energy sources and if "combusted" in fuel cells, the only by-product is water vapor. However, hydrogen has its drawbacks: since it does not exist free in nature, energy must be consumed to extract it from its compounds. It has a high cost (twice the cost of gasoline per energy content) and is difficult to store, especially in an energy-dense form. Table 13-10 compares the properties of various energy carriers.

Hydrogen combusts in air in a much wider range than methane. Its explosion limits are also much wider; it is these properties after all that render it such an interesting fuel. However, hydrogen first goes through a combustion range before going to the explosion range (4% to 13% volume), meaning that it will most probably combust rather than explode, which is not the case for methane (5% to 6%). Hydrogen, being much lighter than air, disperses quickly and much faster than methane.

Hydrogen Production, Storage, and Use

Hydrogen can be produced from water through electrolysis or from fossil fuels through reforming. The energy required by these processes can be obtained from various sources that include fossil fuels, nuclear energy, and renewable energy sources, including biofuels. This plurality in terms of energy sources is one of the main advantages of the hydrogen energy vector, if the world economy can disentangle itself from its dependency on oil.

If hydrogen is produced through the reforming of fossil fuels, then CO_2 is released. Nuclear energy, although CO_2 free, has still to address nuclear waste disposal issues. If hydrogen is produced through water electrolysis, then the emissions related to its production are those associated with the power industry.

TABLE 13-10   Comparison of Combustion Properties of Various Energy Carriers

	Hydrogen	Methane	Propane
LCV (kWh/Nm³)	3	9.9	25.9
Density (kg/m³)	0.09	0.7	2
Concentration for combustion (volume %)	4.1–72.5	5.1–13.5	2.5–9.3
Explosion limits (volume %)	13–65	6.3–14	—
Dispersion coefficient (cm³/s)	0.61	0.15	—

LCV = lower heating value.
The weight of the storage tank for each fuel has not been taken into consideration with the exception of the lead batteries.

Vast quantities of hydrogen as an industrial gas are produced around the world. Total annual production amounts to 500 billion Nm^3/yr, equivalent to less than 10% of the world's oil production in 2002. Almost all of this hydrogen is produced from fossil fuels, as shown in Table 13-11, while only 5% of this hydrogen is commercially used and distributed—the majority is consumed internally in refineries or chemical plants. Commercial hydrogen sales are expected to increase by over 8% per annum till 2008.

Even though electrolysis provides a much more pure form of hydrogen, only a small percent of the global production is obtained in this way in small plants due to the fact that it is much more costly than natural gas reforming, which is three times more energy efficient than electrolysis if fossil source electricity is used (80% for reforming and 40% × 70% = 28% for electricity production and electrolysis). Table 13-12 shows indicative costs for hydrogen production.

The storage of hydrogen is considered its "Achilles' heel." Its storage in an energy-dense form is particularly hard to achieve and is currently one of the many areas of research. Hydrogen is commonly stored in gaseous form under pressure. Large storage tanks are under a pressure of 16 bar while in cylinders hydrogen is stored under 200 to 250 bar. Pressures of 750 bar are being experimentally investigated for applications in the transport sector. Hydrogen in liquid form can be stored in special vacuum tanks, like the ones used in space applications. This type of storage addresses the problem of storing hydrogen at high volume densities; however, liquid hydrogen is still four times less "energy dense" per volume as kerosene. Additionally, 40% of the energy contained in gaseous hydrogen needs to be consumed for lowering the temperature of hydrogen down to 14°Kelvin, where it liquefies.

Innovative storage methods include bonding hydrogen in metal hydrides that are metal dusts whose atom structure allows for the orderly packing of hydrogen atoms, thus achieving higher volume densities than hydrogen in compressed gaseous form (volume is approximately that of gaseous hydrogen at 300 bar, for a tank pressure of 10 bar).

TABLE 13-11 Feedstock Used in the Global Production of Hydrogen

Feedstock	%
Natural gas	48
Oil	30
Coal	18
Electrolysis	4

TABLE 13-12 Hydrogen Production Costs

Method	Cost ($/GJ)
Natural gas reforming	5
Coal gasification	11
Biomass gasification	13
Electrolysis with large-scale hydro	12
Wind electrolysis	32
PV electrolysis	5–100

GJ = gigajoule; PV = photovoltaic.

The weight of these materials is however quite significant, where usually only 1.5% of the total weight is the weight of hydrogen. Depending on the properties of the metallic hydride dust, heat must be supplied to the tank for hydrogen to be released while heat must be absorbed for charging the tanks with hydrogen.

The vast quantities of hydrogen produced today are consumed in non–energy-related uses that are summarized in Table 13-13.

Hydrogen is used in ammonia (NH_3) production that in turn is used for the production of fertilizers. In refineries, hydrogen is used for the upgrading of fuels, mostly for the removal of sulfur. Hydrogen is becoming the single most important product of the refinery so that the final products can meet the ever more stringent fuel specifications; however, it remains an internally consumed product and rarely exits the refinery.

The petrochemical industry uses hydrogen to produce methanol, which is sometimes used in fuel cells where it is reformed to release its hydrogen content. Hydrogen is also used in the food industry for the hydrogenation of fats. Some other uses accrue from the physical properties of hydrogen, like lubrication, heat transfer (cooling of power plant generators), or buoyancy (meteorological balloons).

The space program in the United States has been the only case where hydrogen was used as a fuel. Hydrogen can very well be burned in suitably modified boilers, gas turbines, and internal combustion engines. However, it is the development of fuel cells where hydrogen can be combusted with minimal or no emissions that has opened new horizons for the energetic use of hydrogen in transport, mobile, portable, and stationary applications, spanning all types of human activities.

Hydrogen in the Research Agenda

Driven by recent technical advances in hydrogen and fuel cells technologies and the need for diversified and sustainable technologies, Organisation for Economic Co-operation and Development (OECD) governments are intensifying their research and development efforts. Almost 1 billion euros per year are invested globally for hydrogen and fuel cells research, the three main players being the United States, Japan, and Europe. Half of this amount is spent on fuel cells research and development and the rest on technologies to produce, store, and use hydrogen in other energy-conversion devices like internal combustion engines. The respective investment from the private sector is considerably larger (approximately 3 to 4 billion euros a year), including major oil and gas companies, car manufacturers, electrical utilities, power plant component developers, and a number of "small" players in the current hydrogen and fuel cell market.

Multiannual programs have been announced by the major countries active in the field, including $1.7 billion over 5 years in the United States, 2 billion euros in the 6th Framework Programme and the Growth Initiative of the European Commission, and 30 billion yen per fiscal year in Japan. Similarly significant programs are in place in

TABLE 13-13 Hydrogen Usage

Usage	*%*
Ammonia production	50
Refineries	37
Methanol	8
Space	1
Other	4

(Source: Hart, 1997.)

individual countries like Canada, Germany, Italy, and others. These efforts are complemented by three major international cooperation initiatives:

- The International Energy Agency (IEA) in April 2003 formed the Hydrogen Coordination Group to enhance coordination among national research and development programs, building on the IEA cooperation framework, including the Implementing Agreements on hydrogen, advanced fuel cells, and others.
- In November 2003, 16 countries including non-OECD countries Russia, Brazil, India, and China have formed the International Partnership for the Hydrogen Economy (IPHE), following a proposal of the United States.
- In January 2004, the European Commission established the European Hydrogen and Fuel Cells Technology Platform (IJFP), which is a cluster of public and private initiatives aiming to coordinate and promote the development and application of hydrogen energy technologies including fuel cells.

Hydrogen From Renewables

Long-term forecasts for hydrogen production show some deficiencies between supply and demand, implying that increased production must be covered from alternative energy sources, including renewables. The production of hydrogen from fossil fuels results in CO_2 emissions, the quantities of which per mole of hydrogen produced depend on the feedstock and production technology used.

Natural gas is the fossil fuel with the highest hydrogen-to-carbon ratio; however, hydrogen from electricity-producing renewables or from CO_2-neutral biomass are the ways for producing hydrogen in a distributed fashion without any CO_2 emissions.

Hydrogen production from nuclear energy is also CO_2 free; however, the handling of nuclear wastes is not yet solved while uranium is found in fewer places in the world than oil, meaning that the dependency on few countries possessing raw materials remains.

Also, the technologies used are very complicated and only a few countries and companies possess the knowledge for the development of nuclear plants. Hydrogen from renewables has none of these problems since renewables are indigenous and available around the globe, while RES technologies and hydrogen production and use technologies can be manufactured almost anywhere around the globe.

Hydrogen from fossil fuels can also be CO_2 free if CO_2 sequestration is to be applied. The method has been demonstrated in the North Sea where CO_2 is pumped back into oil wells to enhance the extraction of oil. However, this approach would greatly increase the cost of hydrogen and would create a problem of what to do with the large quantities of the produced CO_2.

So even if fossil and nuclear fuels can be sources of hydrogen in the short to medium term, it is renewables that will be the sources of hydrogen in the long term.

Chapter 14

Microturbine Manufacturing and Packaging

Manufacturing with respect to distributed energy technologies is frequently a euphemism at best. Whether the product in question is a wind turbine or microturbine, the components come from several different suppliers and the "manufacturer" assembles them. The logic of economies of scale must precede any "total ownership of component manufacture" aspirations.

For instance, Solar Turbines in California, a division of Caterpillar, developed microturbine recuperator technology, using a grant from the U.S. Department of Energy (DOE). The DOE has funded several U.S.-based manufacturers for microturbine development work, as described in earlier chapters. Solar Turbines supplies this recuperator to Capstone Turbines, also in California. Capstone assembles its microturbine packages in its Chatsworth, CA, plant using some vendor-supplied parts, such as the recuperator.

Although Capstone has an aggressive marketing division, it extends the effective territory for its products if it partners with other manufacturers, such as Bowman, which is also in California. Bowman packages essentially the same turbine as Capstone can provide. However, both companies can gain from each other's compact size and combined resources, while retaining the advantage of being smaller, and potentially faster on their feet.

Additionally, Capstone partners in many projects, where partners may include branches of government; utilities, both local and foreign; and corporate customers, local and foreign. Different installations therefore may each bring about a different ownership and management structure.

Capstone is doing what other microturbine and other manufacturers of distributed energy products do. The economies of scale demand this endless adaptability. Fortunately, smaller companies traditionally make business moves and changes faster than other companies.

Differing applications (e.g., direct desiccation, recuperation, hybrid) dictate that packaging varies from project to project. Fuel selections (e.g., biodiesel, petrochemical waste, biomass) may demand further customization. Additionally, the assembly-from-components model gives the industry further potential variations on its basic component: the small gas turbine at the heart of the package.

COMPONENT DEVELOPMENT

In summary, component manufacture potential spawns other pools of specialist technologies, fostered by government funding (for instance, Mohawk foil bearings and U.S. DOE grant recipients for their recuperator program) or private money, or both. To further complicate the mix, some companies will develop a component and then offer licenses to others to make that component for the microturbine packagers/manufacturers. Two examples follow: one a discussion of recuperator development and one on foil bearings.

A microturbine is essentially a gas turbine, only smaller than most of us are used to. Although a few technologies will vary, the basic technologies are the same as for conventional gas turbines. Component development, then, is a hard topic to do justice to, in the first edition of a book like this one. The author refers you instead to the upcoming book, *Gas Turbines: A Handbook of Air, Land, and Sea Applications*, by the same author, to be released by Butterworth-Heinemann.

However, there are two key components where the development model for microturbines could vary from their larger counterparts: recuperators (many in established production models) and foil bearings (still at prototype stage). Some work on their development is discussed as follows.

Example 1: Recuperators

Proe 90 Gas Turbine Recuperator[*]

The need for a low-cost and high-performance recuperator has been an obstacle to previous Ericsson Engines (since John Ericsson's original engines). Proe Power Systems has developed a higher-temperature recuperator for the gas turbine industry!

- A totally new heat exchanger construction is shown below. It is not a shell and tube heat exchanger—what you see is what it is.
- Higher turbine exhaust temperatures (815°C/1500°F without superalloys) are allowed, for higher engine efficiency.
- Design innovations enhance creep resistance and long life at elevated temperatures.
- The design avoids thin foils that limit the operating temperature of stainless steel primary surface recuperators to 650°C/1200°F due to creep and corrosion of the fragile foils.
- The system is all welded.
- The construction uses commercially available stainless steel tubing materials and orbital welding techniques developed for shell and tube heat exchangers.
- No special tooling and minimum machine work are required.
- Effectiveness is greater than 90%.
- No complex or costly fin, plate, or primary surface fabrication is involved.
- All welding or brazing involves self-locating parts.

The Proe 90 Gas Turbine Recuperator concept is used in the distributed power microturbine (30 to 50 kW) market.

Current recuperators for gas turbine applications are primary surface heat exchangers that are very expensive to manufacture and typically recover only about 70% to 80% of the exhaust heat. Furthermore, the low creep and corrosion resistance of the thin foils used in conventional primary surface recuperators limit their temperature and restricts the potential engine efficiency. Proe Power Systems' recuperator recovers over 90% of the exhaust heat with minimal exhaust back-pressure.

---

[*] Source: Proe Power Systems. 2002. *Proe 90™ gas turbine recuperator*. Proe Power Systems. Thousand Oaks, CA.

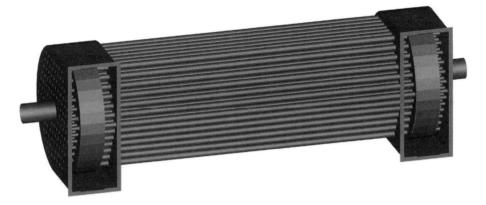

FIGURE 14-1

FIGURE 14-2

Example 2: Foil Bearings*

In terms of rotor load bearing, there are two size categories of microturbine of importance to a manufacturer of foil bearings: the 25- to 75-kW and the 250- to 450-kW range. For both size ranges, eliminating the entire oil lubrication system is a goal that promises low operating and maintenance costs. Mohawk Innovative Technology, Inc. (MiTi) is a world leader in the design and manufacture of high-performance journal and thrust-compliant air foil bearings for use in high-speed rotating machinery.

Current air-bearing manufacturing methods were established to meet the needs of low-volume specialty machinery. Bearing production volumes in the 100- to 300-units per year range dictated the use of low-volume manufacturing processes.

---

* Source: Sayer, J. *Design, development and manufacture of high-speed foil bearings*. Mohawk Innovative Technologies. Albany, NY.

Market projections for microturbines anticipate significant increases in production volume in the coming years. The demand for high-quality compliant foil air bearings is also expected to increase. Current production methods cannot economically provide the volume of air bearings required by the expanding mini- and microturbine market. MiTi has done studies on the potentially high-volume air-bearing market for mini- and microturbines. What follows is a description of their work so far.

Their project objective is:

- To investigate and demonstrate cost-effective, high-volume production methods for compliant foil air bearings
- To establish and demonstrate a cost-effective, high-volume process for verifying assembled bearing quality

MiTi will investigate relevant high-volume manufacturing processes for foil element cutting, bump foil element corrugations forming, and foil element welding. MiTi will then design and demonstrate automated quality-control test stations for high-volume static and dynamic bearing testing. Software will be developed to control test instrumentation and to analyze inspection results for pass/fail/rework criteria.

The availability of high-volume manufacturing and inspection processes will help compliant foil air-bearing technology to compete for application in microturbine distributed generation and to expand the market for MiTi air-bearing products.

Chapter 15

Business Risk and Investment Considerations

This chapter needs to be read in conjunction with all other chapters in Part 2, particularly the chapters that discuss rival sources of distributed power and integrated systems (microturbines in a package with another source of distributed power). Serious investors need to explore all the links and leads these chapters give them as well as all the Securities and Exchange Commission–related documents they can find. Relying on one or even several "one-stop shopping" Internet sites that may charge a high membership fee, that is, an appreciable portion of a microturbine's capital cost, may be ill advised. Two such sites are mentioned in the section titled "Internet Purchase and Internet Paid-Membership Sites" below. However, there is a host of others that include venture capital and investment advice firms. The author will not state that the information these sources provide is not sound, but if one buys it, it ought not be the only decision-making tool one uses.

The sites whose owners are pure finance people are easy to spot. There will be a great deal of *Wall Street Journal* prognoses but very few specifics about which aspects of a multi-partner microturbine project actually made that specific project work. In other words, even if the components of a microturbine are excellent technically, investing in their company or one of their company's projects may not yield profit. In this business sector, *every* case is different (even if some may be similar).

INTERNET PURCHASE AND INTERNET PAID-MEMBERSHIP SITES[*]

Increasingly, the business of small energy and distributed energy sources is being conducted on the Internet. Sometimes the product is for "information only." One information company (Global Information Inc) sells its reports on the Internet. The cost for its microturbine report is listed at just under $4000 U.S.

The microturbine market and the one for fuel cells and its optimum fuel—hydrogen—have a growing population of Internet sites that market, in addition to information, the opportunity to belong to a community listing, where you can post your related products, services, and resume or buy related or advertised products. Fees vary. For instance, http://www.fuelcellmarkets.com as of 2005, charged 5000 British pounds for a membership. In 2006, this was 6850 British pounds for basic services.

For all such Internet sites, one ought to check the conditions and terms of use immediately and every time one thinks about subscribing. This is not a suggestion that such sites are not useful or legitimate. They could save a person a great deal of time if he or she is looking for investment opportunities. However, the truth is that all the

---

[*] Sources: Fuel Cell Markets Portal. Available at: http://www.fuelcellmarkets.com and Global Information Inc's online catalog. Available at: http://www.gii.co.jp.

information that can be found on such sites may be found by individual research using free Internet search engines. The "free" information may also be more accurate and recent. Also, in some cases, private sites may contain less information than one can get from government sources or recipients of government grants for development work, most of which has to be made available to the public.

The following is a partial quote/partial paraphrased extract from the Fuel Cell Markets Web site (2005) that offers this abbreviated explanation of how it came to be:

> At the Lucerne Fuel Cell Conference in July 2001, a number of leading players in the European fuel cell industry [known as the Fuel Cell Europe ad hoc group] got together and concluded that there was [risk] that Europe would [be unprepared] for the commercial introduction of fuel cells, expected worldwide starting around 2005. There was strong support for the establishment of an open industry database to [improve communications and drive developments]. At much the same time, and by fortunate coincidence, Fuel Cell Markets was developing that very system.

It is the responsibility of the individual consumer to investigate all free sources of information before adding to his capital overhead by paying fees to any organization or firm. Most manufacturers of microturbines and microturbine/fuel cell systems will be glad to pass on application case studies (some of which are featured in this book) free of charge.

STOCK OFFERINGS*

Further, when microturbine/distributed-energy companies allow the public to invest in their company, all the information with respect to the stock offering is also made available via the Securities and Exchange Commission and generally on the Internet.

This Internet information needs to be read with caution because it may be released by a third-party company and may not even be reviewed by the manufacturers themselves.

Interestingly, the manufacturers will often have links to these reports on their own Web site but preface the information with a warning about potential inaccuracies (that are caused by the third-party company). Other language referring to the information may include cautions such as: "[Progress] after the publication date and external links may make them [inaccurate]. Historical information [may not be] current and is [for reference only]. [The manufacturer] takes no responsibility for the information contained… [Prognoses can be identified by] words such as *may, will, expect, anticipate, believe, estimate*, and *continue* or similar words. [Prognoses] are subject to numerous [risks] that may cause [the manufacturer's] results to be [different from results implied]."

Risk factors will then be identified in yet another site and need to be studied before purchase. The stock is generally offered as follows:

- Shares of common stock
- Warrants to purchase shares of common stock
- Shares of preferred stock
- Debt securities (senior debt or subordinated debt) or
- Some combination of the above

Although manufacturers "may use agents to sell the above, [they] may decline to. [They] may also decline any purchase." This and similar wording is standard. In today's

---

* Source: Various documents released by the U.S. Securities Exchange Commission on stock offerings from rotating machinery manufacturers.

business climate, this then gives them the option to avoid a hostile takeover if they can see it coming. All agent's fees, earnings, commissions, and discounts are made public for public offerings.

When observing new companies' performance via their Nasdaq stock sticker, note that information regarding technical progress and new innovations that could send the stock value climbing will be available in technical conference proceedings first. Press releases and papers are generally the company's first attempt at product-release advertising.

The best kind of papers to consider when studying the marketability/stock value of technical features are juried papers. Not all conferences jury their papers. A "jury" is a panel of independent (generally unknown to each other and unpaid volunteers) reviewers, generally three in number, who review a paper that has been proposed for presentation at the actual conference. American Society of Mechanical Engineers conferences generally have juried papers.

The American Society of Mechanical Engineers and other similar bodies may also have panel sessions where nothing is published and speakers "present without publication" in the program. These are frequently even more interesting if one gets to dialogue with the speaker(s) at any time. "Without publication" sessions serve as a useful outlet for engineers who may feel obliged to say more than their corporate lawyers will allow them to say in a printed paper.

Conference proceedings are generally available for sale for a nominal fee. $100 U.S. is a typical figure for a CD-ROM. Lists of papers may also be available on the association Web sites and can be bought individually.

In conference papers, pay particular attention to what is *not* said. The only way for a layperson to determine what is "missing" is to compare the work of several manufacturers who make the same category of product. If, for instance, one manufacturer mentions "no leakage at 'x' operating pressure" and OEM B does not, one may wish to flag leakage as a parameter about which to question OEM B, with respect to potential operational flaws.

Potential operational flaws cause rebate, returns, warranty, and other issues that all ultimately reduce earnings on shares. All text of stock offerings will include warnings that caution "operating history is characterized by net losses, and [the manufacturer] anticipates further losses and may never become profitable." This wording is fairly standard for technologies that are relatively new. Some manufacturers of new energy sources may be commercially viable due to government grants that they may receive directly. Alternatively, another manufacturer, or joint venture partner, *A,* may receive government aid to develop the technology or a specific component (say a recuperator on a microturbine) that is then used in the unit that is "manufactured" (packaged) by yet another company, *B.* In turn, *B* may form joint ventures with several other companies that share some combination of manufacture, packaging, or marketing duties and expense.

This can assist essentially the same product to retain a larger geographical supply footprint than if it "went alone." It could also help a company, whose technology is essentially from one country, partner with another company and enter a market region without paying "foreigner surtaxes."

This chain of changing partners may add to risk, however, if component suppliers and their quality control (QC) programs vary. In the large gas turbine world, intermittent (meaning not common to all their licensee) QC problems have plagued machinery fleets in the past, particularly if the model was new.

The current market for microturbines is far from being consistent, let alone sustainable. Wording such as "a sustainable market … may never develop or may take longer [than anticipated], which would adversely affect revenues and profitability" is therefore standard as it prevents the stock seller from being sued for misrepresentation later.

Stock-offering language will therefore consistently point out everything that could legitimately happen, in order to be totally honest. The following cautions are common:

1. "We operate … among competitors who have significantly greater resources than we have and we [might not] compete effectively."
2. "If we do not effectively implement our [business] plans, our sales will [fall] and our profitability will suffer."
3. "We may not be able to retain or develop distributors in our targeted markets, in which case our sales would not increase as expected."

In the case of alternative energy methods, given Europe's higher cultural priorities with respect to alternative energies, item 1 may occur quite commonly. Item 3 is the reason that so many joint ventures occur in the microturbine and hybrid (microturbine with fuel cell) field. However, as already pointed out, QC issues may arise, and the cause of operational problems may be difficult to pinpoint.

With new energy methods, initial customers' success is closely watched by the rest of the target market world. Their success is critical. If they fail, manufacturers/packagers may get blamed for bad service that they did not cause. Overall, market entry could suffer a setback that is reflected in share prices and dividends.

Stock-offering language therefore may also include statements like "… our largest customer's performance as it relates to engineering, installation, and provision of aftermarket services has been below our standards, and, if not rectified, could have a significant impact on our reputation and [stock value]. [If we lose them as a customer], near-term sales, cash flow and profitability could be adversely affected. [This customer also is suffering growth and sales target problems which in turn could affect our] cash flow and profitability targets."

"We may not be able to develop sufficiently trained applications engineering, installation, and service support to serve … targeted markets." This ought to appear in microturbine stock offerings for several years to come. The volume of installations is not adequate to attract excellent service people, who may foresee better job security working for a large gas turbine company/division.

What the reader needs to appreciate here is that the world of microturbines is not yet part of the relatively stable world of conventional large turbines. Something the size of a refrigerator that turns out enough power for your hospital or village to be independent of the local grid represents a new paradigm for the U.S. consumer. That consumer is traditionally less well versed in alternative energies and the pain of their development than their less gas-and-oil endowed European counterparts.

The U.S. business climate for microturbines may be exciting, but it is fraught with many operations considerations that do not occur with a mature gas turbine fleet. So, the manufacturer's warnings in their stock offerings are likely to also include the following:

1. "Changes in product components may require us to replace parts held at distributors and Authorized Service Companies."
2. "[A] highly regulated business environment and changes in regulation could impose costs … or make our products less economical, thereby affecting [product] demand."
3. "Utility companies or governmental entities could place barriers to [market entry] and we may not be able to effectively sell our product."
4. "Product quality expectations may not be met, causing slower market acceptance or warranty cost exposure."
5. "[T]he development of new products and enhancements of existing products [dictate our business success]."

6. "Operational restructuring may result in asset [reduction] or other unanticipated charges."
7. "[The] production cost reductions necessary to competitively price product [may not occur], which would impair our sales."
8. "Commodity market factors impact costs and [materials] availability."
9. "[S]uppliers may not supply us with a sufficient amount of components or components of adequate quality...."
10. "[Due to] a lengthy sales cycle, and sales [projections], our potential profitability [may suffer]."
11. "Potential intellectual property, stockholder, or other litigation may adversely impact [profit margins]."
12. "[F]unding for future operating requirements may not be forthcoming, which could force us to curtail operations or close."
13. "We may not manage growth, expand production, or improve operational, financial, and management information systems [to a degree that ensures] sales and profitability."
14. "[S]uccess depends in significant part upon the service of management and key employees [that we may not be able to retain]."
15. "[F]uture effectiveness of internal controls over financial reporting or the impact on operations or stock price is uncertain."
16. "[P]otentially significant fluctuations in operating results, and the market price of our common stock ... may change regardless of our operating performance."

For companies operating in California, they have to also allow for earthquakes and earthquake insurance. For companies operating in the southeast United States, they also have to allow (increasingly) for flood and hurricane damage.

Any stock offering will also include (but not be limited to) basic information on the following:

- The offering company
- The company's product
- How proceeds from the stock offering will be used
- Financial performance/earnings to date

The ratio of earnings to fixed charges may be presented on a quarterly basis if the company is relatively new and if significant change has occurred in its last year of operation.

The description of stock will include the number of total shares, shares outstanding, and stock value. Typically, new companies start with low stock value to attract investors and hope that the stock will climb.

Bylaws will spell out provisions and safeguards for the company and its investors against hostile takeover attempts and other terms of stock ownership. Different states have different anti-takeover legislation, so that needs to be studied. There are also different provisions with respect to stockholders who own a large amount of stock. Timing here is critical. If the stockholder bought the stock before certain regulations were in effect, then that ownership is not governed by the new regulations.

GOVERNMENT SUPPORT AND NEW LEGISLATION

Key considerations to investment in a high capital, longer return-on-investment business, include but are not limited to the following:

- Does the company being considered have its government's support/backing?
- Which government is involved—federal or state?
- Is the government assistance of a tax rebate and/or research and development funds nature?
- Is the government agency itself doing the research and development or paying the private companies to do it, or both?
- How much money has the government given and over how many years?
- Does current legislation suggest that government funding will continue?
- Will funding stop due to short-term politics (e.g., a Republican versus a Democratic government)?
- What kind of lobby does the machinery type/energy type have in Washington?
- What kind of competition/synergy/joint ventures can the global industry offer? How mature is the industry?
- What is the manufacturer's capability, given the above, to reduce the research and development and manufacturing/service costs?

With respect to the last two items above, for instance, with wind energy: several joint U.S.–European manufacturer/operator projects exist and share resources. If the U.S. federal government were to withhold production tax credits from a U.S. plant, the business would suffer, but it would survive. The microturbine, even given the aggressive joint venturing efforts of a company such as Capstone Turbine, is not yet stable in its place in global energy.

When studying government support, the reports on progress so far with the U.S. Department of Energy (DOE) are available on the Internet. The extract pertaining to microturbines is quoted below.

The U.S. Department of Energy's (DOE's) Distributed Energy Program[*] is working with utilities, energy service companies, industrial manufacturers, and equipment suppliers to identify technologies that will improve the energy, environmental, and financial performance of power systems for manufacturing, processing, and other commercial applications. The program will contribute to the development of ultrahigh-efficiency and low-emission engine systems and provide new choices and innovative power solutions to the industrial sector. The Advanced Microturbine Systems Program Plan for Fiscal Years 2000 Through 2006 outlines proposed activities to develop advanced microturbine systems for distributed energy applications.

- Advanced microturbine program
- Testing and validation
- Materials program
- Simulator

Advanced Microturbine Program

The Advanced Microturbine Program run by the U.S. DOE is a 6-year program for fiscal years 2000 to 2006 with a government investment of more than $60 million. End-use applications for the program are open and include stationary power applications in industrial, commercial, and institutional sectors. The program includes competitive solicitations for engine conceptual design, development, and demonstration; component, subsystem, and system development; and development of a technology base in the

---

[*] Source: U.S. Department of Energy, Energy Efficiency and Renewable Energy, Distributed Energy Program—Microturbines.

areas of materials, combustion, and sensors and controls. Technology evaluations and demonstrations are also part of the program.

Planned activities focus on the following performance targets for the next generation of "ultra-clean, high-efficiency" microturbine product designs:

- High efficiency: Fuel-to-electricity conversion efficiency of at least 40%
- Environment: Nitric oxides < 7 parts per million (natural gas)
- Durability: 11,000 hours of reliable operations between major overhauls and a service life of at least 45,000 hours
- Cost of power: System costs < $500/kW, costs of electricity that are competitive with alternatives (including grid) for market applications
- Fuel flexibility: Options for using multiple fuels including diesel, ethanol, landfill gas, and biofuels

The five manufacturers involved in the program are:

- Capstone
- General Electric
- Ingersoll-Rand
- Solar Turbines
- United Technologies

Testing and Validation

One of the supporting elements of the program is the testing and validation of microturbines at the University of California–Irvine (UCI) Distributed Technologies Testing Facility. Southern California Edison, in partnership with UCI, is leading this project. This $2.1 million project, which was started in 1996, receives co-funding from the California Energy Commission and the Electric Power Research Institute. The project's goal is to determine the availability, operability, reliability, and performance characteristics of commercially available microturbines. It will compare manufacturer claims with actual installation, operation, and testing of units and assess microturbine performance against South Coast Air Quality Management District emissions rules and Institute of Electrical and Electronics Engineers power quality standards.

Materials Program

New materials such as combustion liners and high-temperature material recuperators are designed and tested to endure and perform properly in microturbine-specific environments. A jump in microturbine efficiency can be achieved through increases in engine operating temperatures. Ceramics and metallic alloys are being developed to accomplish this. For additional information about microturbine materials, see the Microturbine Materials Technology Activities presentation.

The materials projects listed below are led by Oak Ridge National Laboratory (ORNL). More detailed information about each of these projects is available in the following reports:

- Recuperators
- Ceramics and composites
- Coatings, power electronics, and advanced renewable energy systems (ARES).

Recuperators

- Creep Behavior of Advanced Alloys for High-Temperature (660°C–750°C) Microturbine Recuperators (ORNL)
- Materials Selection for High-Temperature (750°C–1000°C) Metallic Recuperators for Improved Efficiency Microturbines (ORNL)
- Composition Optimization for Corrosion Resistance to High-Temperature Exhaust Gas Environments (ORNL)
- Microturbine Recuperator Testing and Evaluation (ORNL)

Monolithic Ceramics

- Hot Section Components in Advanced Microturbines (Honeywell Ceramic Components)
- Oxidation/Corrosion Characterization of Microturbine Materials (ORNL)
- Mechanical Evaluation of Monolithic Ceramics Containing Environmental Barrier Coatings (ORNL)
- Hot Section Materials Characterization (University of Dayton, Research Institute)
- Reliability Evaluation of Microturbine Components (ORNL)
- Oxidation-Resistant Coatings on Silicon Nitride for Microturbines (ORNL)
- Advanced Surface Treatments of Silicon Nitride (ORNL)

Ceramics Life Prediction

- NDE Technology Development for Microturbines (Argonne National Laboratory)
- Reliability Analysis of Microturbine Components (Connecticut Reserve Technologies)

Power Electronics

- Development of High-Efficiency Carbon Foam Heat Sinks for Microturbine Power Electronics (ORNL)

Simulator

Another element of the DOE's efforts is the microturbine simulator project. It is a co-development effort with DOE's energy-storage activities, the National Rural Electric Cooperative Association, and the Electric Power Research Institute. The project is developing a simulator/model to mimic actual performance characteristics of a microturbine and validate simulators at actual utility sites using data from commercial microturbines. In addition, the project will install simulators to test peak-shaving, power-quality, baseload, and other applications. For more information, read the report on the microturbine simulator demonstration.

Investors also need to study the research and devlopment budget figures allocated to development work. How large is the budget? More importantly, how large are other energy sources' budgets? What does this say about government commitment to this energy sector?

What other governments are doing work in this sector? What is the likely synergy from these other studies?

After reading the information on government support, a potential investor ought to also read relevant sections of the new U.S. energy bill. Microturbines get a 10% rebate under that act. How did other energy sources fare? Is commitment to microturbines demonstrated sufficiently? These are the questions one needs to ask.

CULTURAL CONSIDERATIONS*

Cultures vary across continents, countries, states, and regions. Just as people have ethnic cultures, governments also have their unique cultures with different emphases on items such as environmental and pollution priorities. For instance, California will add state grants to federal grants and give photovoltaics a better return-on-investment period, at least as far as current legislation goes. Will that last? What will change for microturbines with respect to tax incentives? In which states? If those are known, can the volume of foreseeable business justify investment?

Other questions the investor needs to ask are as follows: Do European assets/programs/business volume for the company in question justify investment? Are there other international assets/programs? What is the growth trend of the microturbine industry elsewhere that can be supplied by the (U.S. or otherwise) supplier?

Additionally, what is the potential for this energy source in concert with other energy sources? The microturbine with a fuel cell has achieved 80% efficiency. Is this combination well received? Is it slated to become widespread?

It ultimately does not matter how many advantages an energy source offers. If the consumer is staid and does not feel comfortable using the equipment or "being seen as an owner" of one, it does not matter how good the equipment is. Ultimately, to flourish on a large scale, microturbines need household customers. Their small- to medium-sized business operators are not likely to create enough business volume on their own. So it comes down to the household or, rather, the head of the household. Maybe a microturbine just does not fit his or her image of "sexy" or "manly" or "safe." Like the Hummer's and SUV sales prove, energy conservation does not matter much to the U.S. citizen, even in this day and age. Does he or she think a wind turbine or photovoltaic system or solitary fuel cell better fits his or her concept of a household functioning properly? Or is he or she so mobile in this age of job insecurity that the local utility is the only viable option?

FEDERAL LIFE-CYCLE COSTING PROCEDURES

Federal agencies are required to evaluate energy-related investments on the basis of minimum life-cycle costs (10 CFR Part 436). A life-cycle cost evaluation computes the total long-run costs of a number of potential actions and selects the action that minimizes the long-run costs. When considering retrofits, sticking with the existing equipment is one potential action, often called the *baseline* condition. The life-cycle cost (LCC) of a potential investment is the present value of all of the costs associated with the investment over time.

The first step in calculating the LCC is the identification of the costs. *Installed cost* includes the cost of materials purchased and the labor required to install them (e.g., the price of an energy-efficient lighting fixture, plus the cost of labor to install it). *Energy cost* includes annual expenditures on energy to operate equipment (e.g., a lighting fixture that draws 100 watts and operates 2000 hours annually requires 200,000 watt-hours [200 kWh] annually; at an electricity price of $0.10 per kWh, this fixture has an annual energy cost of $20). *Non-fuel operations and maintenance* includes annual expenditures on parts and activities required to operate equipment (e.g., replacing burned-out lightbulbs). *Replacement costs* include expenditures to replace equipment upon failure (e.g., replacing an oil furnace when it is no longer usable).

---

* Source: Articles written by Claire Soares for various Power Generation journals, 2004 and 2005.

Because LCC includes the cost of money, periodic and aperiodic maintenance (O&M), and equipment-replacement costs, energy-escalation rates, and salvage value, it is usually expressed as a present value, which is evaluated by:

$$LCC = PV(IC) + PV(EC) + PV(OM) + PV(REP)$$

where *PV(x)* denotes "present value of cost stream *x*," *IC* is the installed cost, *EC* is the annual energy cost, *OM* is the annual non-energy O&M cost, and *REP* is the future replacement cost.

Net present value (NPV) is the difference between the LCCs of two investment alternatives, e.g., the LCC of an energy-saving or energy-cost-reducing alternative and the LCC of the existing, or baseline, equipment. If the alternative's LCC is less than the baseline's LCC, the alternative is said to have a positive NPV, i.e., it is cost-effective. NPV is thus given by:

$$NPV = PV(EC_0) - PV(EC_1)) + PV(OM_0) - PV(OM_1)) + PV(REP_0) - PV(REP_1)) - PV(IC) \text{ or}$$

$$NPV = PV(ECS) + PV(OMS) + PV(REPS) - PV(IC)$$

Where *subscript 0* denotes the existing or baseline condition, *subscript 1* denotes the energy cost-saving measure, *IC* is the installation cost of the alternative (note that the IC of the baseline is assumed zero), *ECS* is the annual energy cost savings, *OMS* is the annual non-energy O&M savings, and *REPS* is the future replacement savings.

Levelized energy cost (LEC) is the break-even energy price (blended) at which a conservation, efficiency, renewable, or fuel-switching measure becomes cost-effective (NPV ≥ 0). Thus, a project's LEC is given by:

$$PV(LEC*EUS) = PV(OMS) + PV(REPS) - PV(IC)$$

where *EUS* is the annual energy use savings (energy units per year). Savings-to-investment ratio (SIR) is the total (PV) savings of a measure divided by its installation cost:

$$SIR = (PV[ECS] + PV[OMS] + PV[REPS])/PV(IC)$$

Chapter 16

The Future for Microturbine Technology

Where is the ultimate future of microturbines headed? Several perspectives include palm-sized or smaller units that can provide all the power a household needs. Technology's evolution is indeed taking us in that approximate direction.

As is evident from the material in this chapter, much of the technology to get to "personal pocket-sized gas turbines" is already here. Now the main quasi help/hindrance to the "individual's own power plant" is the big oil/power/corporation model that currently rules global business. None of us may live to see that model allow some of the technology described in this chapter become mainstream, but one never knows.

Stationary gas turbines got to be the sophisticated items they are courtesy of technology that filtered down from the National Aeronautics and Space Administration to military aviation to commercial aviation to land-based and marine engines. It is now quite common, for instance, to see triple redundancy controls in power-generation units. Not that long ago, the term *triple redundancy* conjured up images of a military fighter jet.

Because the U.S. military always wants an edge, they are spending the money to develop the microturbines of the future. After the companies they fund to produce these cutting-edge propulsion units for their latest weapons or spy machines have finished their development work, they will look for more funding. If there is no need to give them any more, the company will seek commercial consumer markets to make a living. At that point, the microturbine "PT" (personal turbine) could become reality in a hurry.

Consider the following material from M-Dot Aerospace[*] that receives U.S. government—particularly U.S. military—funding. This company specializes in the design of small gas turbines and turbomachinery, including the following:

- Turbojets for unmanned vehicles and microturbojets to power micro air vehicles
- Turboprops for unmanned vehicles (heavy fuel engines)
- Microturbines
- Small turbofans

With U.S. government (e.g., Defense Advanced Research Projects Agency [DARPA], Army, Navy, and Air Force) funding, the company has developed a 94-horsepower twin spool turboprop for unmanned vehicles, called the TPR80. The engine design can be produced in recuperated and non-recuperated form. They are developing a microturbine with DARPA funding that will propel micro air vehicles. This engine can be configured as a variety of turbomachines including microturbojet, turboprop, or turboshaft configuration.

---

[*] Source: M-Dot Aerospace Web site. Available at: http://www.m-dot.com.

M-Dot Aerospace's work in the fields of ultra-high-speed turbine, alternator, and electric motor development is aimed at speeds in excess of 500,000 RPM. One million RPM is their target for a microturbine, motor, or alternator.

The company's other turbomachines and systems include the following:

- Small turbopumps for rockets and missiles
- Gas turbine exhaust noise suppressors and mufflers for aircraft and APUs
- Tiny centrifugal compressors, microcompressors, and tiny blowers
- Micro refrigerators and coolers
- Exhaust eductors and ejectors
- Stainless steel manifolds

They also developed a tiny jet fuel turbopump for the Navy capable of achieving 1300 psi output pressure at roughly 100,000 RPM.

Let us now consider the "peacetime" applications that may be developed by a university, such as Massachusetts Institute of Technology (MIT). Universities have a multinational force of cheap "whiz kids" (i.e., research students) with no security clearance, so they seek funding for peacetime ventures.

Their work also brings us closer to the personal turbine of the future.

"THUMBNAIL" -SIZED PERSONAL TURBINES[*]

By harnessing silicon microfabrication techniques, engineers hope to build gas-turbine engines weighing just 1 gram by the turn of the century.

In 1959, in a speech at the California Institute of Technology in Pasadena, Nobel Prize–winning physicist Richard Feynman presented an intuitive perspective of the concept of micromachines, exploring greatly reduced forces on tiny parts, proportionally greater strength of materials, and so on.

More than 40 years later, engineers at the Gas Turbine Laboratory of MIT in Cambridge, MA, are addressing the same technical issues that Feynman raised. The MIT group is working on a gas-turbine engine that can fit on a dime (Figure 16-1).

FIGURE 16-1 *Measuring just 4 millimeters in diameter, this radial inflow turbine wheel was manufactured from silicon using deep reactive ion etching. (Source: Ashley, S. 1997. Turbines on a dime. ASME Magazine. The American Society of Mechanical Engineers. New York, NY.)*

[*] Reference: Ashley, S. 1997. Turbines on a dime. *ASME Magazine.* The American Society of Mechanical Engineers. New York, NY.

The entire device—complete with an integrated electric generator—is expected to weigh 1 gram. The researchers have already manufactured a 4-millimeter-diameter radial inflow turbine wheel from silicon using deep reactive ion etching—a relatively new microfabrication method. With minor changes in the airfoil shapes, the same component will function as a centrifugal compressor wheel as well. Work continues on a suitable combustor unit.

Calculations indicate that this prototype can be made of a temperature-resistant material like silicon carbide. A complete gas-turbine generator system—with a volume less than 1 cubic centimeter—could deliver as much as 50 watts of electric power.

The benefits of such technical development effort are obvious. The greatest potential for changing everyday life as we know it is portable power production, a personal turbine or "PT" for the average consumer once mass production lowered costs enough. The energy density of liquid hydrocarbon fuel is 20 to 30 times that of the best battery technology, so the possibility exists that these power sources could shrink proportionally.

The turbine would likely consume less than 10 grams/hour of hydrogen fuel, according to the MIT team's estimates. The fuel supply, an exhaust system, and presumably a thermally insulating containment vessel would add to size and weight. Several microturbines might operate in parallel, perhaps in arrays developed using wafer-scale integration techniques.

The team has studied the basic fundamentals; developed a baseline design; and performed the necessary research on components, materials, and fabrication methods. Component testing, detailed design, and initial engine fabrication are next.

The microturbine project presents new challenges in the mechanical and electrical engineering disciplines of fluid dynamics, structural mechanics, bearings and rotor dynamics, combustion, and electric machinery design.

Downscaling Issues

Physics and mechanics influencing the design of the components do change with scale. Therefore, the optimal detailed designs can be quite different. Examples of these scaling effects include the viscous forces in the fluid (which are larger at microscale), surface-area-to-volume ratios (also larger at microscale), chemical reaction times (invariant), the electric-field strength that can be realized (higher at microscale), and manufacturing constraints (limited mainly to two-dimensional planar geometries).

The group in the Gas Turbine Laboratory has performed a careful scaling study of high-speed, rotating turbomachinery, which has shown that suitably designed microdevices are remarkably promising.

The requirements for many power-production applications favor a larger engine size: 50 to 100 watts. Viscous effects in the fluid flow and combustor residence time requirements also favor larger engine size. However, semiconductor manufacturing technology sets size limits; the upper size limit is set mainly by the etching-depth capability, which is currently a few hundred microns, while the lower limit is set by feature resolution and aspect ratio capabilities of the production processes.

Miniaturization

The success of the miniature heat engine requires the development of three microscale technologies:

- Rotating machinery
- Combustor technology
- High-temperature material fabrication technology

Design calculations indicate that the device, to have sufficient power per unit weight, will require:

- Combustor exit temperatures of 1000°C to 1500°C
- Rotor peripheral speeds of 300 to 600 meters/second
- Rotating structures centrifugally stressed to several hundred megapascals
- Low friction bearings
- High dimensional tolerances
- Thermal isolation of the hot and cold sections

High values of power per unit weight depend on high tip speed because the power transferred to the air follows the square of the peripheral speed. Therefore, the microdevice has to turn at the same peripheral speed as a large turbine. High speed implies high centrifugal stress because stress rises with the square of the rotational speed.

In conventional practice, rotor speed is constrained to several hundred meters/second by the strength-to-density ratio of high-temperature metal alloys. However, defect-free microfabricated materials are quite strong. Materials at microscale are much better than at macroscale, so with single-crystal silicon, structures that would be unfeasible using standard metals such as titanium or superalloys can be designed.

In addition, their density is only half that of superalloys. So compared with macroscopic materials, these nonmetallic materials have a superior strength-to-density ratio, to allow them to spin at high speeds without the risk of fracture.

High-temperature performance of silicon is limited by its creep life, so a more refractory material is needed for the practical microturbine effort. Work is under way at the laboratory to design and fabricate silicon carbide and silicon/silicon carbide hybrid structures by chemical vapor deposition of relatively thick silicon carbide layers (10 to 200 microns) over silicon molds.

High rotating speeds need low-friction bearings. Here the cubic scaling of volume and mass combined with the quadratic scaling of areas (the so-called cube-square law) mean that the surface-area–to–weight ratio of a rotor is large at small scales. This implies that miniature air bearings can support large loads.

Rotating Disks

The baseline microturbine looks like a single-spool turbojet.

The highly integrated design consists of a supersonic radial flow compressor and turbine connected by a hollow shaft (to limit heat conduction). The rotor is 12 millimeters in diameter and 3 millimeters long. During operation, the wheels would spin at 2.5 million RPM.

Upon start-up, gaseous hydrogen fuel is injected at the compressor exit and mixes with air as it flows radially outward to the flame holders. The combustor discharges radially inward to the turbine whose exhaust turns 90 degrees to exit the engine nozzle. The baseline engine may have a pressure ratio of 4:1 and airflow of 0.15 grams/second. A thin-film electric starter/generator will mount on a shroud over the compressor blades and be cooled by compressor discharge air. Compressor discharge air will also cool the structure to isolate the compressor thermally from the combustor and turbine. The rotor is supported on air bearings.

The motor-driven compressor has similar geometry to that of the turbine, with the airfoils modified to those of a radial outflow compressor and the electrical machine configured as a motor. The design application in this case is the pressurization of 100-watt fuel cells. As an engine is scaled down linearly, its power-to-weight ratio increases.

The development of a small-scale combustor seemed problematic initially because chemical reaction times are invariant with size and the tiny volume of the device provides limited flow residence time. So the MIT team has successfully demonstrated a 2-millimeter-long combustion chamber that is 40 times the relative size of a conventional combustor.

The baseline generator design, meanwhile, is a 500-pole planar electric induction machine mounted on the shroud of the compressor rotor. The use of an electrostatic generator unit, as opposed to an electromagnetic one, was necessary because at macroscales, an electromagnetic device has much greater power density.

Microfabrication

Researchers plan to demonstrate the manufacture of all the microengine components in silicon first, although they are developing silicon carbide microfabrication methods.

A deep reactive ion etching produced the 4-millimeter turbine wheel, which features blades with spans of 200 microns, and a 10-micron bearing gap between the rotor disk and the stator base plate.

If the power per unit of airflow remains constant (for the baseline engine, about 130 watts/gram/second), then indications are that a millimeter-size engine would have a thrust-to-weight ratio of about 100:1, compared with 10:1 for the best modern aircraft engines.

References

Government, Industry, and Renewable Energy Web Sites

GOVERNMENT AND NON-PROFIT ORGANIZATIONS

National Renewable Energy Laboratories: http://www.nrel.gov
National Wind Technology Center: http://www.nrel.gov/wind/index.html
U.S. Department of Energy: http://www.eren.doe.gov
Iowa Energy Center: http://www.energy.iastate.edu
Iowa Department of Natural Resources: http://www.state.ia.us/government/dnr
Renewable Energy Policy Project: http://www.repp.org
Izaak Walton League of America: http://www.iwla.org
Sandia National Laboratories: http://www.sandia.gov

WIND ENERGY

American Wind Energy Association: http://www.awea.org

SOLAR ENERGY

Solar Energy Laboratory: http://sel.me.wisc.edu
Solar Energy Industries Association: http://www.seia.org

BIODIESEL

National Biodiesel Board: http://www.biodiesel.org
Biodiesel Information Centre: http://www.greenfuels.org/bioindex.html
U.S. Department of Energy's National Biofuels Program:
http://www.biofuels.doe.gov
Western Biomass: http://www.westbioenergy.org/March1999/index.htm
Biofuels Information Center: http://www.afdc.nrel.gov

GASIFICATION

Gasification Technologies Council:
http://www.gasification.org/story/story.html
Caddet Renewable Energy: http://www.caddet-re.org

ANAEROBIC DIGESTION

American Bioenergy Association: http://www.biomass.org/sites-andig.htm
AgSTAR: http://www.epa.gov/agstar

MICRO-HYDROELECTRIC

U.S. Department of Energy Study:
http://www.eren.doe.gov/consumerinfo/refbriefs/ab2.html
Energy Systems and Design (Low Head Micro-Turbines):
http://www.microhydropower.com
Oasis Montana Inc. (Aquair Submersible hydro generator):
http://www.oasismontana.com/AQUAIR.html
Canyon Industries Inc. (Measuring Head and Flows):
http://www.geocities.com/~canyon5
Independent Power & Light (cost break down micro-hydro):
http://www.independent-power.com/small_hydro1.html
http://www.independent-power.com/Harris1-6.htm
http://www.independent-power.com/microhydro1.html

BATTERY STORAGE SYSTEMS

Northern Arizona Wind & Sun, Inc:
http://windsun.com/Batteries/Deep_Cycle.htm
Backwoods Solar Electric Systems: http://www.backwoodssolar.com
Independent Power & Light: http://www.independentpower.
com/medium_system.htm
Go Solar Company (Battery distributor):
http://www.solarexpert.com/Catbattery.html

PUMPED HYDRO

Integrative Science: Renewable Energy:
http://community.hei.com/altenergy/hydro.pshydro.html
Pumped Storage Hydro Station:
http://193.86.119.10/elektrarny_en/vodni/dlouhe_strane/html/ke_dlouhe_strane.htm

FLYWHEELS

Regenerative Power and Motion: http://rpm2.8k.com/homepage.htm

FUEL CELLS

Fuel Cells 2000: http://216.51.18.233/index_e.html
U.S. Department of Energy, Fuel Cells and Transportation:
http://www.ott.doe.gov/oaat/fuelcell.html
Department of Defense Fuel Cell Web Links:
http://dodfuelcell.com/helpfulsites.html

MOTHER EARTH NEWS

http://www.motherearthnews.com/altenergy/altenergy176.cells.middle.html
Hydrogen Fuel Cell Program: http://www.dri.edu/Projects/Energy
Energy Partners: http://www.energypartners.org/product.htm

MICROTURBINES

Gas Research Institute: http://www.gri.org/cgibin/
re?url=http%3A//www.gri.org/pub/content/jan/20000117/143950/micro_4.html

MODERN POWER SYSTEMS

http://www.capstoneturbine.com/press/Power899.htm
Global Gas Turbine News: http://asme.org/igti/ggtn/archives.html
Electrical Generating System Association:
http://www.egsa.org/powrline/nd98/98nd_micro.htm

STIRLING ENGINES

Sunpower Inc.: http://www.sunpower.com
American Stirling Company: http://www.stirlingengine.com
Whisper Tech Ltd.: http://www.whispertech.co.nz
Stirling Engine Web Site: http://www.bekkoame.ne.jp/~khirata/english/howwork.htm
Stirling Thermal Motors: http://www.stmcorporation.com

GOVERNMENT INCENTIVES

U.S. Department of Energy Million Solar Roofs Program:
http://www.eren.doe.gov/millionroofs
Wind Powering America Initiative:
http://www.eren.doe.gov/windpoweringamerica
Iowa Alternative Energy Revolving Loan Program:
http://www.energy.iastate.edu/about/grantloan/AERLP
http://www.energy.iastate.edu/renewable/AssistDev.html
Iowa Energy Bank: http://www.state.ia.us/dnr/energy/programs
Rebuild Iowa: http://www.state.ia.us/dnr/energy/programs

Index